Methods in Molecular Biology

Series Editor
John M. Walker
School of Life and Medical Sciences
University of Hertfordshire
Hatfield, Hertfordshire, UK

For further volumes:
http://www.springer.com/series/7651

For over 35 years, biological scientists have come to rely on the research protocols and methodologies in the critically acclaimed *Methods in Molecular Biology* series. The series was the first to introduce the step-by-step protocols approach that has become the standard in all biomedical protocol publishing. Each protocol is provided in readily-reproducible step-by-step fashion, opening with an introductory overview, a list of the materials and reagents needed to complete the experiment, and followed by a detailed procedure that is supported with a helpful notes section offering tips and tricks of the trade as well as troubleshooting advice. These hallmark features were introduced by series editor Dr. John Walker and constitute the key ingredient in each and every volume of the *Methods in Molecular Biology* series. Tested and trusted, comprehensive and reliable, all protocols from the series are indexed in PubMed.

Cereal Genomics

Methods and Protocols

Edited by

Luis M. Vaschetto

Instituto de Diversidad y Ecología Animal, Consejo Nacional de Investigaciones Científicas y Técnicas (IDEA, CONICET), Universidad Nacional de Córdoba, Córdoba, Argentina; Agronomy, Horticulture and Plant Science Department, South Dakota State University, Brookings, SD, USA

Humana Press

Editor
Luis M. Vaschetto
Instituto de Diversidad y Ecología Animal, Consejo Nacional de
Investigaciones Científicas y Técnicas (IDEA, CONICET)
Universidad Nacional de Córdoba
Córdoba, Argentina

Agronomy, Horticulture and Plant Science Department
South Dakota State University
Brookings, SD, USA

ISSN 1064-3745 ISSN 1940-6029 (electronic)
Methods in Molecular Biology
ISBN 978-1-4939-9864-7 ISBN 978-1-4939-9865-4 (eBook)
https://doi.org/10.1007/978-1-4939-9865-4

This Humana imprint is published by the registered company Springer Science+Business Media, LLC part of Springer Nature.
The registered company address is: 233 Spring Street, New York, NY 10013, U.S.A.

Preface

The cereal grasses (Poaceae) are a fundamental source of nutrients in the human diet, and major cereal crops, which include wheat, maize, rice, barley, oat, sorghum, and millet, represent primary food sources for populations around the world. In the last years, research efforts in molecular biology have been aimed at increasing not only yield and grain quality in cereal crops but also their adaptability to ever-changing environments. Nowadays, the emergence of high-throughput sequencing technologies, along with bioinformatic tools to analyze big volumes of data and genome-editing platforms to dissect gene pathways, has modified the mode of working with genetic information and complex traits. In the next years, it is expected that both basic and applied research in cereal genomics will play a crucial role to achieve sustainable global food security.

This volume of Methods in Molecular Biology is the result of the collaboration of leading scientists from around the world (including Argentina, Australia, Canada, China, France, Germany, India, Japan, Mexico, the United Kingdom, and the United States). Here, I have compiled some of the latest molecular techniques for the analysis and manipulation of grass genomes. Cutting-edge technologies and advancements in cereal genomics are covered as follows: high-throughput DNA extraction protocols, crop genetic resources, meta-quantitative trait loci (QTL) analysis, association mapping, next-generation genome sequencing, transposable element-associated variation, transcriptomics analysis, epigenetic variation (i.e., both DNA methylation and histone modifications), identification of imprinted genes, noncoding RNAs (circular RNAs), genome-editing technologies, and posttranslational protein phosphorylation. Each chapter has carefully been selected so that the reader can explore the hottest topics and apply current protocols in cereal genomics.

Córdoba, Argentina
Brookings, SD, USA

Luis M. Vaschetto

Contents

Contributors

KEVIN BEGCY • *Environmental Horticulture Department, University of Florida, Gainesville, FL, USA*

PANKAJ BHOWMIK • *National Research Council Canada, Saskatoon, SK, Canada*

ANDRIY BILICHAK • *Morden Research and Development Center, Agriculture and Agri-Food Canada, Morden, MB, Canada*

CHEN CHEN • *Jiangsu Key Laboratory of Crop Genetics and Physiology, Co-Innovation Center for Modern Production Technology of Grain Crops, Key Laboratory of Plant Functional Genomics of the Ministry of Education, Yangzhou University, Yangzhou, China*

HUW R. DAVIS • *Institute of Biological, Environmental and Rural Sciences (IBERS), Aberystwyth University, Aberystwyth, UK*

ANURAG DAWARE • *National Institute of Plant Genome Research (NIPGR), Aruna Asaf Ali Marg, New Delhi, India*

CLELIA DE-LA-PEÑA • *Unidad de Biotecnología, Centro de Investigación Científica de Yucatán, Mérida, Yucatán, Mexico*

THOMAS DRESSELHAUS • *Cell Biology and Plant Biochemistry, Regensburg Center for Biochemistry, University of Regensburg, Regensburg, Germany*

FÁTIMA DUARTE-AKÉ • *Unidad de Biotecnología, Centro de Investigación Científica de Yucatán, Mérida, Yucatán, Mexico*

JUNJI EMURA • *Peptide Institute, Inc., Osaka, Japan*

ALISON M. R. FERRIE • *National Research Council Canada, Saskatoon, SK, Canada*

MARÍA PAULA FILIPPONE • *Universidad Nacional de Tucumán, Facultad de Agronomía y Zootecnia, (UNT-FAZ), Tucumán, Argentina*

TSUYOSHI FURUMOTO • *Department of Agriculture, Ryukoku University, Shiga, Japan*

DANIEL GAUDET • *The University of Lethbridge, Lethbridge, AB, Canada*

ROCÍO LILIANA GÓMEZ • *Instituto de Tecnología Agroindustrial del Noroeste Argentino (ITANOA), Estación Experimental Agroindustrial Obispo Colombres (EEAOC)—Consejo Nacional de Investigaciones Científicas y Técnicas (CONICET), Tucumán, Argentina*

ELSA GÓNGORA-CASTILLO • *CONACYT-Unidad de Biotecnología, Centro de Investigación Científica de Yucatán, Mérida, Yucatán, Mexico*

PRATEEK GUPTA • *Repository of Tomato Genomics Resources, Department of Plant Sciences, University of Hyderabad, Hyderabad, India*

MARCELO HELGUERA • *Laboratorio de Biotecnología EEA INTA Marcos Juárez, Marcos Juárez, Argentina*

ROBERT J. HENRY • *Queensland Alliance for Agriculture and Food Innovation, University of Queensland, Brisbane, QLD, Australia*

KATSURA IZUI • *Institute of Advanced Technology, Kindai University, Wakayama, Japan*

HUW D. JONES • *Institute of Biological, Environmental and Rural Sciences (IBERS), Aberystwyth University, Aberystwyth, UK*

Sateesh Kagale • *National Research Council Canada, Saskatoon, SK, Canada*
Taiji Kawakatsu • *Institute of Agrobiological Sciences, National Agriculture and Food Research Organization, Tsukuba, Ibaraki, Japan*
John Laurie • *Agriculture and Agri-Food Canada, Lethbridge, AB, Canada*
Laura Levin • *Davison Life Sciences, University of Georgia, Athens, GA, USA*
Damon Lisch • *Department of Botany and Plant Pathology, Purdue University, West Lafayette, IN, USA; Center for Plant Biology, Purdue University, West Lafayette, IN, USA*
Celso Gaspar Litholdo Jr • *Centre National pour la Recherche Scientifique (CNRS)/ Université de Perpignan Via Domitia (UPVD)—Laboratoire Génome et Développement des Plantes (LGDP—UMR5096), Perpignan, France*
Zhanjie Li • *Key Laboratory of Genetics, Breeding and Multiple Utilization of Crops, Ministry of Education, Fujian Provincial Key Laboratory of Haixia Applied Plant Systems Biology, Fujian Agriculture and Forestry University, Fuzhou, China; National Engineering Research Center of Sugarcane, Fujian Agriculture and Forestry University, Fuzhou, China*
Anne L. Maddison • *Institute of Biological, Environmental and Rural Sciences (IBERS), Aberystwyth University, Aberystwyth, UK*
Asuka Nishimura • *Laboratory of Plant Molecular Genetics, Graduate School of Agricultural and Life Science, The University of Tokyo, Tokyo, Japan*
Swarup K. Parida • *National Institute of Plant Genome Research (NIPGR), Aruna Asaf Ali Marg, New Delhi, India*
Dylan W. Phillips • *Institute of Biological, Environmental and Rural Sciences (IBERS), Aberystwyth University, Aberystwyth, UK*
Nandhakishore Rajagopalan • *National Research Council Canada, Saskatoon, SK, Canada*
William A. Ricci • *Davison Life Sciences, University of Georgia, Athens, GA, USA*
Hymavathi Salava • *Repository of Tomato Genomics Resources, Department of Plant Sciences, University of Hyderabad, Hyderabad, India*
Lorena Noelia Sendín • *Instituto de Tecnología Agroindustrial del Noroeste Argentino (ITANOA), Estación Experimental Agroindustrial Obispo Colombres (EEAOC)—Consejo Nacional de Investigaciones Científicas y Técnicas (CONICET), Tucumán, Argentina*
Rameshwar Sharma • *Repository of Tomato Genomics Resources, Department of Plant Sciences, University of Hyderabad, Hyderabad, India*
Cynthia G. Soto-Cardinault • *Unidad de Biotecnología, Centro de Investigación Científica de Yucatán, Mérida, Yucatán, Mexico*
Yellamaraju Sreelakshmi • *Repository of Tomato Genomics Resources, Department of Plant Sciences, University of Hyderabad, Hyderabad, India*
Claudia Mabel Terenti Romero • *Instituto Nacional de Tecnología Agropecuaria, Estación Experimental Agropecuaria San Luis (INTA, EEA SAN LUIS), San Luis, Argentina*
Akhilesh K. Tyagi • *National Institute of Plant Genome Research (NIPGR), Aruna Asaf Ali Marg, New Delhi, India; Department of Plant Molecular Biology, University of Delhi South Campus, New Delhi, India*
Yoshihisa Ueno • *Department of Agriculture, Ryukoku University, Shiga, Japan*
Tomoko Urabe • *Peptide Institute, Inc., Osaka, Japan*

LUIS M. VASCHETTO • *Instituto de Diversidad y Ecología Animal, Consejo Nacional de Investigaciones Científicas y Técnicas (IDEA, CONICET), Universidad Nacional de Córdoba, Córdoba, Argentina; Facultad de Ciencias Exactas, Físicas y Naturales, Universidad Nacional de Córdoba, (FCEFyN, UNC), Córdoba, Argentina; Agronomy, Horticulture and Plant Science Department, South Dakota State University, Brookings, SD, USA*

KAI WANG • *Key Laboratory of Genetics, Breeding and Multiple Utilization of Crops, Ministry of Education, Fujian Provincial Key Laboratory of Haixia Applied Plant Systems Biology, Fujian Agriculture and Forestry University, Fuzhou, China; National Engineering Research Center of Sugarcane, Fujian Agriculture and Forestry University, Fuzhou, China*

TAKU YOSHIYA • *Peptide Institute, Inc., Osaka, Japan*

KUMIKO YOSHIZAWA-KUMAGAYE • *Peptide Institute, Inc., Osaka, Japan*

XIAOYU ZHANG • *Davison Life Sciences, University of Georgia, Athens, GA, USA*

XINYAN ZHANG • *Department of Botany and Plant Pathology, Purdue University, West Lafayette, IN, USA*

MEIXIA ZHAO • *Department of Botany and Plant Pathology, Purdue University, West Lafayette, IN, USA*

Chapter 1

A Low-Cost High-Throughput Method for Plant Genomic DNA Isolation

Prateek Gupta, Hymavathi Salava, Yellamaraju Sreelakshmi, and Rameshwar Sharma

Abstract

Many of the functional genomics methods require isolation of genomic DNA from large population of plants. The selection of DNA isolation protocols depends on several factors such as choice of starting material, ease of handling, time and labor required for isolation, the final quantity as well as the quality of genomic DNA. We outline here a high-throughput method of DNA extraction from different plant species including cereal crops. The protocol can be used for extraction of DNA in single tubes as well as for large formats in 96-well plates. The protocol includes steps for eliminating interfering secondary products such as phenolics. This protocol can be applied for high-throughput isolation of DNA for varied applications such as TILLING, mapping, fingerprinting, etc. as a cost-effective protocol compared to commercial kits.

Key words DNA isolation, Genomic DNA, Corn, Maize, Cereals

1 Introduction

The advent of functional genomics has necessitated genomic DNA isolation from a large number of plants. Due to the presence of rigid non-cellulose components in the plant cell wall of cereal crops, DNA isolation requires mechanical and efficient homogenization of tissues to release the DNA from the cells. In addition, homogenization releases metabolites such as nucleases, carbohydrates, chlorophylls, and secondary metabolites, which interfere with DNA isolation and PCR amplifications. Moreover, the presence of polysaccharides, proteins, and DNA polymerase inhibitors such as tannins, alkaloids, and polyphenols makes it difficult for the extraction and purification of high-quality DNA from cereals [1]. Dellaporta et al. [2] reported an efficient Miniprep protocol for the isolation of genomic DNA from maize tissues. This protocol has been widely adapted for many plant species; however, in several species the presence of contaminating metabolites, especially secondary metabolites such as polyphenolics, phenolics, and flavonoids, can influence

Luis M. Vaschetto (ed.), *Cereal Genomics: Methods and Protocols*, Methods in Molecular Biology, vol. 2072,
https://doi.org/10.1007/978-1-4939-9865-4_1,

DNA quality. Efficient removal of these contaminants during DNA isolation is essential, as these metabolites bind to DNA imparting brown color and interfere with the downstream PCR amplifications [3]. Xin and Chen [4] reported a DNA isolation method using CTAB with additional cleanup step from sorghum leaves and seeds. Most protocols enabling isolation of genomic DNA from plants possessing high levels of polyphenols use a phenol-chloroform extraction step [5, 6]. However, this step necessitates phase separation and careful collection of the aqueous phase, which precludes use of these protocols for high-throughput DNA isolation. Leach et al. [7] used a urea-based extraction buffer for high-throughput isolation of DNA from maize leaves. Recently, Abdel-Latif and Osman [1] compared different DNA isolation methods for the isolation of genomic DNA from maize grains and found that the modified Mericon method (Qiagen DNeasy Mericon Kit) was more efficient in terms of yield and quality.

High-throughput genomics applications such as gene/QTL mapping, insertional mutagenesis, and TILLING require isolation of genomic DNA from large population. This necessitates a low-cost and high-throughput protocol for DNA isolation. However, the presence of secondary metabolites, such as polyphenols in tomato [8], entails inclusion of a step to efficiently eliminate these metabolites in early stage of extraction. The inclusion of polymers such as polyvinylpolypyrrolidone (PVPP) in extraction buffer can reduce the binding of polyphenols to DNA [6, 8, 9]. The addition of β-mercaptoethanol can reduce the activity of peroxidases and polyphenol oxidases, thus preventing oxidation of the polyphenols [10]. The incubation of genomic DNA with RNase for 30–60 min can remove the contaminating RNA [6, 11]. The inclusion of PVPP, β-mercaptoethanol, and RNase during initial extraction steps yielded high-quality genomic DNA for TILLING from tomato cotyledons [12]. In this chapter, we outline a low-cost high-throughput DNA extraction method, which can be applied to any plant tissue. Using this protocol genomic DNA from about 384 plant samples (four 96 deep well plates) can be isolated in a single run.

2 Materials

2.1 Equipment

1. Deep well plate tissue homogenizer (Mini Beadbeater, BioSpec Products Inc.)
2. Centrifuge with microplate rotor (SH-3000, Sorvall Evolution RC, Thermo Scientific).
3. Multichannel pipettes or 96-channel Pipettor (Eppendorf/PP550 DS, Apricot Designs Inc.)
4. Nanodrop Spectrophotometer (ND-1000, Thermo Fisher Scientific).

5. Temperature controlled Water Bath or Incubator (Daihan Labtech Co.)
6. Horizontal gel electrophoresis system (Genei).
7. Gel documentation system (Syngene).

2.2 DNA Extraction

Prepare all solutions with molecular biology grade chemicals, in sterile nuclease-free MilliQ water. Autoclaving of all plastic wares and other accessories is essential.

1. DNA Extraction buffer: 0.1 M Tris–HCl (pH 7.5), 0.05 M EDTA (pH 8.0), 1.25% (w/v) SDS.
2. Polyvinylpolypyrrolidone (PVPP) (Sigma-Aldrich).
3. β-mercaptoethanol (Sigma-Aldrich).
4. 10 mg/mL RNase A (Sigma-Aldrich).
5. 6 M Ammonium acetate (Sigma-Aldrich).
6. Isopropanol (SRL).
7. Ethanol (70% v/v) (Hayman).
8. TE buffer: 10 mM Tris (pH 7.5), 1 mM EDTA (pH 8 0.0).
9. 96-well plates (2 mL and 1 mL) (Axygen Scientific).
10. Steel balls of ~2 mm diameter.
11. Forceps and spatula.

2.3 DNA Quantification, Dilution, and Storage

1. 96-well PCR plates (Axygen Scientific).
2. Silicon sealing mats (Axygen Scientific).
3. Tris buffer: 10 mM Tris (pH-7.5).

3 Methods

3.1 DNA Isolation

The protocol for DNA extraction in a 96-well plate from maize and wheat leaf tissue is outlined below. This protocol can also be used for DNA extraction in microcentrifuge tubes with suitable modifications (*see* **Note 1**). The usage of a 96-channel pipettor or an 8- or 12-channel pipette is recommended for pipetting to save time. At one time, four 96-well plates can be processed for DNA extraction, subject to the availability of a swinging bucket rotor capable of accommodating four deep well plates.

1. Harvest about 100 mg of fresh leaf and place it in the 96 deep well plate kept on the ice (*see* **Notes 1–3**).
2. Place three steel balls using forceps into each well (*see* **Note 4**). In parallel also add 20 mg of insoluble PVPP into each well (*see* **Note 5**).

3. Using a multichannel or a 96-channel pipettor, add 750 μl preheated (65 °C) extraction buffer to each well. Add 10 μl β-mercaptoethanol to a final concentration of 0.2 M to each well of the plate (*see* **Note 6**).
4. Reseal the plates tightly with silicon sealing mats and homogenize the tissue in a Mini-Bead Beater for 2 min (*see* **Note 7**).
5. Incubate the plates in a water bath or in an oven at 65 °C for 30 min for cell lysis. Mix the contents of plates at about 5–7 min intervals by gently inverting the plates.
6. Transfer the plates to room temperature for 5 min and add 4 μL of 10 mg/mL RNAse A to each well. Reseal the plates and gently mix the contents and incubate them at 37 °C in a water bath or incubator for 30 min.
7. Transfer the plates to room temperature for 5 min and add 400 μL of ice-cold ammonium acetate to each well and mix thoroughly by gentle inversion. Incubate the plates at 4 °C for 15 min.
8. Centrifuge the plates at 4750 × *g* for 30 min at 4 °C in a swinging bucket rotor (*see* **Note 8**).
9. Transfer 650–700 μL of the clear supernatant using a multichannel or a 96-channel pipettor to a fresh 2 mL 96-well plate (*see* **Note 9**).
10. Add an equal volume of ice-cold isopropanol to each well to precipitate DNA. Mix gently and incubate at −20 °C for at least 2 h (*see* **Note 10**).
11. Centrifuge the plates at 4750 × *g* for 30 min at 4 °C. Remove the supernatant by gentle inversion of the plate.
12. Add 500 μL of ethanol (70% v/v) to each well to wash the pellet. Spin at 4750 × *g* for 10 min at 4 °C and discard the supernatant.
13. Repeat the ethanol wash step. Incubate the plates at 65 °C for 10–15 min to dry the DNA pellet (*see* **Note 11**).
14. Add 100 μL TE buffer to each well and store plates at 4 °C overnight to dissolve DNA (*see* **Note 12**).
15. Centrifuge the plates at 4750 × *g* for 30 min at 4 °C to pellet any undissolved impurities present in DNA.
16. Transfer about 90 μL of the supernatant to a fresh 1 mL 96-well plate. Seal with a fresh sealing mat and store the plates at −20 °C.

3.2 Quantification, Dilution, and Storage

Since repeated freezing and thawing can degrade DNA, it is recommended that DNA is diluted immediately after extraction and aliquoted for desired genomic applications. The steps outlining the dilution and aliquoting of DNA are described below:

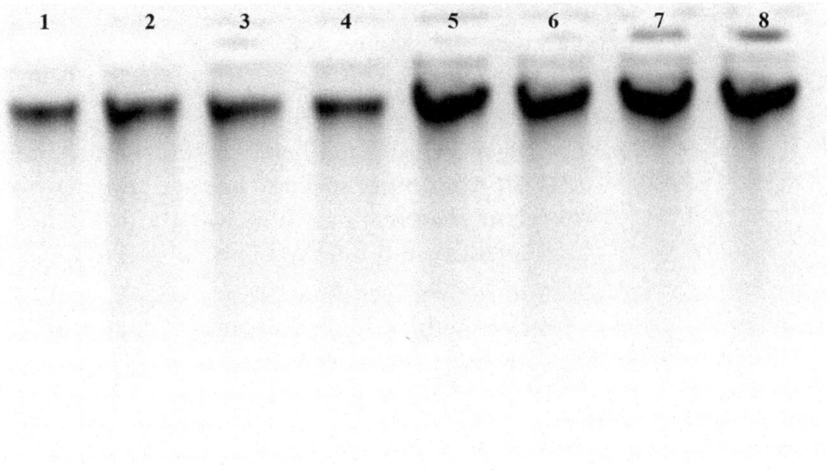

Fig. 1 The profile of genomic DNA isolated using described protocol. Lanes 1–4: Genomic DNA isolated from maize leaves and Lanes 5–8 from wheat leaves. The Nanodrop data for the samples are listed below: Lane 1 is 415 ng/μL (A_{260}/A_{280}—2.03, A_{260}/A_{230}—2.22), Lane 2 is 670 ng/μL (A_{260}/A_{280}—2.01, A_{260}/A_{230}—2.11), Lane 3 is 511 ng/μL (A_{260}/A_{280}—2.00, A_{260}/A_{230}—2.17), Lane 4 is 330 ng/μL (A_{260}/A_{280}—2.02, A_{260}/A_{230}—2.22), Lane 5 is 642 ng/μL (A_{260}/A_{280}—2.05, A_{260}/A_{230}—2.16), Lane 6 is 1400 ng/μL (A_{260}/A_{280}—2.10, A_{260}/A_{230}—2.27), Lane 7 is 1344 ng/μL (A_{260}/A_{280}—2.04, A_{260}/A_{230}—2.16), and Lane 8 is 1078 ng/μL (A_{260}/A_{280}—2.01, A_{260}/A_{230}—2.19)

1. Incubate the DNA samples at 65 °C for at least 15 min for uniform dissolution of DNA and to avoid any stratification of DNA before quantification.
2. Estimate the quantity and quality of DNA by electrophoresis in 1% (w/v) agarose gel containing 0.5 μg/mL ethidium bromide (Fig. 1). Make sure that the DNA is intact. In addition, the quality and quantity of DNA can be examined by a spectrophotometer (*see* **Note 13**).
3. Equalize all the DNA samples to a 5 ng/μL concentration using Tris buffer. Use this plate as a master plate for the DNA stock.
4. From this plate, transfer 5 μL aliquots to each well of a fresh 96-well PCR plate using a 96-channel pipettor.
5. Dry the PCR plates containing 5 ng DNA per well by centrifugation in a vacuum drier for 8–10 min. Cover the plates with aluminum sealing mats and store at −20 °C until required.

4 Notes

1. If the number of samples is less, the extraction can be done in microcentrifuge tubes (1.5–2 mL). For DNA extraction in a 1.5 mL microcentrifuge tube, 100 mg of tissue with same volume of the solution described above can be used. The extraction process can be scaled up or down depending on the requirement using vials/plates of appropriate volume.

2. For efficient isolation of DNA, the total fresh weight of tissue placed in a single well of the plate should not exceed 80–100 mg. The tissue can be collected using a leaf punch yielding tissue of desired weight.
3. The protocol outlined here can also be used for the isolation of DNA from other species. The age and growth conditions of the plant material can influence the extraction efficiency of high quantity and quality DNA.
4. We routinely use commercially available steel balls used in fly-wheel of the bicycle. We discard them after single use.
5. Since PVPP is insoluble, it has to be added separately to each well of the plate. To facilitate equal addition to each well, a scoop can be designed to draw the desired weight of PVPP. The amount of PVPP added should be optimized as it is specific to species to eliminate polyphenol contamination. Wear a mask to avoid inhalation. If the tissue does not have any polyphenols, it can be avoided.
6. β-mercaptoethanol should be freshly added to each well of the plates. Carry out addition of β-mercaptoethanol in a fume hood. The use of β-mercaptoethanol during homogenization also reduces polyphenol contamination in the extracted DNA.
7. Ensure tight fitting of sealing mat to the plate before beginning homogenization, to avoid cross-contamination between wells. The leakage if any due to sealing mat can be tested in a mock experiment by adding a colored dye to the extraction buffer in few wells and its leakage to neighboring well can be monitored after homogenization. During homogenization, some fragmentation of DNA will happen, which causes slight smear on the gel. The duration of tissue homogenization can be optimized by examining the homogenate under a dissection microscope. Excessive homogenization of tissue though may improve DNA yield; it also leads to the shearing of DNA.
8. This step removes the precipitated proteins, polysaccharides, and other debris. In case of deep well plate, the maximum centrifugation speed is limited to about 4750 × g (Sorvall, SH-3000 rotor). The centrifugation at higher speeds at this step and later steps, which can be used for microcentrifuge tubes, helps in removing the impurities more efficiently and hence improves the quality of DNA.
9. While using a 96-channel pipettor for this step, pre-standardize the level for pipette tips so that the required volume can be drawn from the wells without disturbing the debris at the bottom.
10. Isopropanol is found to be more efficient for precipitation of high molecular weight DNA. Compared to ethanol, less volume of isopropanol is required for DNA precipitation. It also separates the polysaccharides. In case of samples with low DNA

yield, an overnight incubation after addition of isopropanol helps in increasing the yield. For saving time, DNA precipitation can also be done at −80 °C with 1-h incubation.

11. Make sure that no trace amount of ethanol is left since ethanol can inhibit downstream reactions like PCR. Avoid overdrying of the DNA pellet. Slow drying of DNA pellets can be done at room temperature too.
12. Supplementing TE buffer with 3.2 μg/mL RNase increases the storage life of DNA.
13. Spectrophotometric measurement can be done using a Nanodrop spectrophotometer with 1 μL of the sample. It is essential to ascertain that the isolated DNA is free of major contaminants such as polysaccharides, polyphenols, and RNA. This can be judged by clear color, A_{260}/A_{280} ratio (1.8–2.0), and A_{260}/A_{230} ratio (>1.5).

Acknowledgments

The development of this protocol was supported by the Department of Biotechnology, New Delhi, India.

References

1. Abdel-Latif A, Osman G (2017) Comparison of three genomic DNA extraction methods to obtain high DNA quality from maize. Plant Methods 13:1
2. Dellaporta SL, Wood J, Hicks JB (1983) A plant DNA minipreparation: version II. Plant Mol Biol Rep 1:19–21
3. Angeles JGC, Laurena AC, Tecson-Mendoza EM (2005) Extraction of genomic DNA from the lipid-, polysaccharide-, and polyphenol-rich coconut (*Cocos nucifera* L.). Plant Mol Biol Rep 23:297–298
4. Xin Z, Chen J (2012) A high throughput DNA extraction method with high yield and quality. Plant Methods 8:26
5. Fulton TM, Chunwongse J, Tanksley SD (1995) Microprep protocol for extraction of DNA from tomato and other herbaceous plants. Plant Mol Biol Rep 13:207–209
6. Porebski S, Bailey LG, Baum BR (1997) Modification of a CTAB DNA extraction protocol for plants containing high polysaccharide and polyphenol components. Plant Mol Biol Rep 15:8–15
7. Leach KA, McSteen PC, Braun DM (2016) Genomic DNA isolation from maize (Zea mays) leaves using a simple, high-throughput protocol. Curr Protoc Plant Biol 1:15–27
8. Pich U, Schubert I (1993) Midiprep method for isolation of DNA from plants with a high content of polyphenolics. Nucleic Acids Res 21:3328
9. Hong YK, Kim SD, Polne-Fuller M et al (1995) DNA extraction conditions from *Porphyra perforata* using LiCl. J Appl Phycol 7:101–107
10. Kim CS, Lee CH, Shin JS et al (1997) A simple and rapid method for isolation of high quality genomic DNA from fruit trees and conifers using PVP. Nucleic Acids Res 25:1085–1086
11. Tel-Zur N, Abbo S, Myslabodski D et al (1999) Modified CTAB procedure for DNA isolation from epiphytic cacti of the genera Hylocereus and Selenicereus (Cactaceae). Plant Mol Biol Rep 17:249–254
12. Sreelakshmi Y, Gupta S, Bodanapu R et al (2010) NEATTILL: a simplified procedure for nucleic acid extraction from arrayed tissue for TILLING and other high-throughput reverse genetic applications. Plant Methods 6:3

Chapter 2

Cereal Genomics Databases and Plant Genetic Resources in Crop Improvement

Robert J. Henry

Abstract

Cereal improvement is based upon effective utilization of genetic resources. These include germplasm and genomics data and tools. Cereal germplasm is available from major global seed banks. Wild material remains an additional less well utilized resource. Sourcing of germplasm requires protocols to ensure intellectual property matters are adequately addressed. Advances in genomics technology have made extensive data set available for the cereals. Reference genome sequences, transcriptome resources, and pan genomes are now available for the major cereal species. The use of genomic data is facilitated by the addition of user-friendly interfaces that allow breeders to access the information they need.

Key words DNA sequences, Germplasm, Seed banks, Gene banks

1 Introduction

Rapid advances in DNA sequencing technology have enabled an explosion of genomic data on all organisms including cereals [1]. This has created a need for improved approaches to the storage of data and the development of tools for data access and analysis. Genetic resource collections have a continuing need to ensure that their limited resources are used effectively to conserve the diversity of cereal species and make them readily available to support cereal genetic improvement. Genomics can provide tools for better characterization of germplasm and may facilitate access to appropriate genetic resources. However, this requires systems to allow accessions in gene banks to be easily linked to genomic data and the resources to collect that data.

2 Materials

Genetic resource material is usually obtained under a Material Transfer Agreement (MTA) for exchange of genetic material or

Luis M. Vaschetto (ed.), *Cereal Genomics: Methods and Protocols*, Methods in Molecular Biology, vol. 2072, https://doi.org/10.1007/978-1-4939-9865-4_2,

associated data (*see* **Note 1**). It may also be necessary to obtain a permit to collect material in the wild. This will vary depending upon local laws and the exact site of collection. For example, material collected on private land may have a different status to that obtained in a protected area such as a national park.

3 Methods

Methods for accessing plant genetic resources and data are described below. Access to germplasm is governed by complex arrangements that depend upon the laws of the countries involved and international treaties and agreements as ratified.

3.1 Addressing IP Issues

For plant genetic resources the first step toward access is to ensure all intellectual property issues are addressed. The Access and Benefit Sharing Clearinghouse (http://absch.cbd.int) provides information on how to obtain permission to access plant genetic resources at the country level. A standard material transfer agreement should be used (*see* **Note 1**).

One or more of the following agreements or laws may control access.

3.2 CBD

The Convention on Biological Diversity (CBD) dates from 1993. This international agreement empowers countries to create laws to control ownership of their biological resources.

3.3 FAO Treaty

The International Treaty on Plant Genetic Resources for Food and Agriculture (FAO Treaty) controls access to 65 major crops with multilateral benefit sharing mechanisms (http://www.fao.org/plant-treaty/en/) and entered into force in 2004. The major cereal crops are listed under the FAO treaty but many minor crops are not. For the cereals, *Hordeum, Oryza, Pennisetum, Secale, Sorghum, Triticosecale, Triticum,* and *Zea* are included but other related genera are not.

3.4 Nagoya Protocol

The Nagoya Protocol on Access and to Genetic Resources and the Fair and Equitable Sharing of Benefits Arising from their Utilization or Nagoya Protocol (NP) came into effect in 2014 with ratification by the required number of countries. The NP requires countries to establish local laws to ensure appropriate access and benefit sharing.

3.5 Local Laws

Local laws control ownership and access in specific cases. In federations these may be at the state rather than national level. For example, in Australia, some states have laws that control access to biological resources within that state.

3.6 PBR (UPOV)

Plant Breeders Rights (PBR) provides for the protection of intellectual properties rights of plant breeders and vary with country but are respected internationally through the International Union for the Protection of New Varieties of Plants (UPOV). Access to protected genotypes for commercial uses may require an agreement with the owner of the PBR.

3.7 Patents

In some cases plants or plant genes may be protected by traditional patents. This will depend very much on the specific geographical location of the use including use of the products derived from the plants. Cereals produced for export of grain to another country may also need to be compliant with rights in the country of destination.

3.8 Sources of Germplasm

Genetic resources for cereals may be found in gene banks or seed banks, in wild populations, or in cultivation in fields cultivated by farmers. These diverse sources are all important for accessing different types of diversity.

3.9 Resources in Seed Banks

Very large numbers of accessions are available for the key crop species (Table 1). Resources in seed banks can generally be obtained using a Standard Material Transfer Agreement (SMTA) (*see* **Note 1**) as defined under the FAO treaty. Many of these seed banks contain accessions that have been sourced from other collections so the total number of unique accessions available is difficult to determine.

3.10 Wild Resources

Wild resources may represent a more diverse set of traits than that easily obtained for a seed bank. These are more difficult to access but may provide novel genetic variation. Genetic resources in wild populations can be utilized by following the guidelines above to obtain access. Specific permits to collect for either research or commercial purposes may be required by local authorities. It may be useful to obtain accessions collected from these populations in established gene banks. This avoids the significant cost of field collection and may also simplify management of IP issues.

Table 1
Cereal collections available from gene banks [2]

Species	Common name	Number of accessions
Triticum aestivum	Wheat	856,168
Oryza sativa	Rice	773,948
Hordeum vulgare	Barley	466,531
Zea mays	Maize	327,932
Sorghum bicolor	Sorghum	235,688
Avena sativa	Oat	130,653

Table 2
Cereal genomics resources

Species	Common name	Reference	Databases
Triticum aestivum	Wheat	International Wheat Genome Sequencing Consortium [3]	
Oryza sativa	Rice	3010 diverse cultivated rice [4] 13 *Oryza* species [5]	
Hordeum vulgare	Barley	[6]	
Zea mays	Maize	[7]	
Sorghum bicolor	Sorghum	[8]	

3.11 Resources in Cultivation

Farmers' fields and commercial seed and breeding collections also represent an important source of germplasm. Access to these resources requires specific arrangements with the owners of the seed.

3.12 Sequence Databases

The Genebank sequence database includes extensive publically available sequence data. These databases also allow comparison of cereal gene sequences with those of other organisms. Reference genome sequences are available for major species (Table 2). These reference sequences have been progressively updated as sequencing technology has improved. Analysis of diversity within species is provided by data on sequence variation for specific genotypes. This is often referred to as re-sequencing data. Transcriptome sequence databases provide a resource for use in analysis of gene expression in specific tissues at particular developmental stages or in specific genotypes. The concept of the pan genome has emerged as a way of considering all of the variation in the genome of a species allowing for assessment of the extent to which some genes may be present or absent in different genotypes. Data may be represented as the invariant part of the genome (present in all genotypes) and the variable part (present or absent in different genotypes).

3.13 Short Read Databases

Genome sequences generate large quantities of short DNA sequences. Databases to store this are a source of this short read data. A Sequence Read Archive (SRA) also known as the short read archive was established by the International Nucleotide Sequence Database Collaboration (INSDC) and stores this data for public use.

3.14 Pan Genome Resources

Pan genomes are available for cereal species [9]. This extends the data available for high-quality reference genomes to provide information on the components of the genome that vary between genotypes within the species. For example, in rice (*Oryza sativa*), analysis of more than 1400 genotypes revealed more than 8000 genes that were not present in the reference genome based upon the cultivar, Nipponbare [10].

3.15 Phenotypic Databases

Data on plant phenotypes is available in a range of databases. These are usually species specific. The great diversity of phenotypes that may be of interest to users especially plant breeders makes this an important area for innovation in automation of data capture.

3.16 Cereal-Specific Databases

1. Wheat: A reference genome for hexaploid wheat [11] is available as are reference genomes for the diploid progenitor species [12]. A fully annotated high-quality genome of wheat cultivar Chinese Spring with 107,891 genes has been produced by the International Wheat Genome Sequencing Consortium (2018). The International Wheat Genome Sequencing Consortium has also established an online resource (https://wheat-urgi.versailles.inra.fr/Seq-Repository) for access to wheat genome resources [13].

   Transcriptomic data from diverse wheat genotypes has been reported for the developing grain [14]. This allows analysis of genes that influence wheat functional and nutritional quality.

2. Rice: Rice genomic resources include high-quality reference genomes for the *Oryza* species [5] and re-sequencing data for many domesticated varieties and wild accessions. Extensive data on the rice transcriptome is also available.

   The Rice Genome Annotation project (http://rice.plantbiology.msu.edu/) provides an annotated genome sequence of each of the 12 rice chromosomes for the reference genotype, Nipponbare [15]. Extensive transcriptome data is also available from this site.

   A rice pan genome is available that incorporates data from a large number of genotypes [16].

   Oryzabase (https://shigen.nig.ac.jp/rice/oryzabase/) is a database for genetic and phenotypic data on rice.

3. Maize: The maize genome has been extensively studied and extensive resources are available. The pan genome concept has been important in understanding the variation in the genomes of maize genotypes.

4. Barley: Barley genome sequences have been reported for wild and domesticated genotypes. A high-quality reference genome is available for barley cultivar Morex [6].

5. Sorghum: The reference sorghum genome is complemented by data on other genotypes.

   The reference genome has been updated several times [8].

QTLs are available in an atlas (https://aussorgm.org.au/) that provides a key resource for linking the sorghum genome with traits for plant breeding.

6. Oat: A transcriptome resource has been produced for oat [17] and reference genomes are being developed.

4 Note

1. Use of material requires a Standard Material Transfer Agreement between those providing and receiving material under the Multilateral System. This agreement may not be varied or abbreviated in any way. It only requires the details of the parties to be added to the template agreement.

References

1. Henry RJ (2017) Plant genetic resources. In: Hunter D, Guarino L, Spillane C, McKeown P (eds) Routledge handbook of agricultural biodiversity. Routledge, London, pp 15–29
2. Wambugu P, Ndjiondjop M-N, Henry RJ (2018) Role of genomics in promoting the utilization of plant genetic resources in genebanks. Brief Funct Genomics 17:198–206
3. International Wheat Genome Sequencing Consortium (2018) Shifting the limits in wheat research and breeding using a fully annotated reference genome. Science 361:eaar7191
4. Wang W, Mauleon R, Hu Z et al (2018) Genomic variation in 3010 diverse accessions of Asian cultivated rice. Nature 557:43–49
5. Stein JC, Yu Y, Copetti D et al (2018) Genomes of 13 domesticated and wild rice relatives highlight genetic conservation, turnover and innovation across the genus *Oryza*. Nat Genet 50(2):285–296. https://doi.org/10.1038/s41588-018-0040-0
6. Mascher M, Gundlach H, Himmelbach A et al (2017) A chromosome confirmation capture ordered sequence of the barley genome. Nature 544:427–436
7. Jiao Y, Peluso P, Shi J et al (2017) Improved maize reference genome with single-molecule technologies. Nature 546:524–527
8. McCormick RF, Truong SK, Sreedasyam A et al (2018) The *Sorghum bicolor* reference genome: improved assembly, gene annotations, a transcriptome atlas and signatures of genome organization. Plant J 93:338–354
9. Golicz AA, Batley J, Edwards D (2016) Towards plant pangenomics. Plant Biotechnol J 14:1099–1105
10. Yao W, Li GW, Zhao H et al (2015) Exploring the rice dispensable genome using a metagenome-like assembly strategy. Genome Biol 16:187
11. Zimin AV, Puiu D, Hall R et al (2017) The first near-complete assembly of the hexaploid bread wheat genome, *Triticum aestivum*. Gigascience 6:1–7. https://doi.org/10.1093/gigascience/gix097
12. Luo MC, Gu YQ, Puiu D et al (2017) Genome sequence of the progenitor of the wheat D genome *Aegilops tauschii*. Nature 551:498–502
13. Alaux M, Rogers J, Letellier T et al (2018) Linking the International Wheat Genome Sequencing Consortium bread wheat reference genome sequence to wheat genetic and phenomic data. Genome Biol 19:111
14. Rangan P, Furtado A, Henry RJ (2017) The transcriptome of the developing grain: a resource for understanding seed development and the molecular control of the functional and nutritional properties of wheat. BMC Genomics 18:766
15. Kawahara Y, de la Bastide M, Hamilton JP et al (2013) Improvement of the Oryza sativa Nipponbare reference genome using next generation sequence and optical map data. Rice 6:4
16. Zhao Q, Fen Q, Lu H et al (2018) Pan-genome analysis highlights the extent of genomic variation in cultivated and wild rice. Nat Genet 50:278–284
17. Gutierrez-Gonzalez JJ, Tu ZJ, Gravin DF (2013) Analysis and annotation of the hexaploid oat seed transcriptome. BMC Genomics 14:471

Chapter 3

Integrated Genomic Strategies for Cereal Genetic Enhancement: Combining QTL and Association Mapping

Anurag Daware, Swarup K. Parida, and Akhilesh K. Tyagi

Abstract

Identification of genetic basis for important agronomic traits is essential for marker-assisted crop improvement. Linkage mapping is one of the most popular approaches utilized for identification of major quantitative trait loci (QTLs) governing important agronomic traits in cereals. However, the identified QTLs usually span large genomic intervals and very few of these are subsequently fine mapped to single major effect gene. This hinders application of these QTLs in marker-aided breeding and crop genetic enhancement. On the contrary, association mapping, another popular approach for identification of QTLs, provides very high resolution but suffers from high level of false positives. Joint linkage-association analysis provides a way to combine advantages and avoid the pitfalls associated with both these methods. In this context, we recently developed MetaQTL specific regional association analysis and demonstrated its utility to rapidly narrow down previously identified QTL intervals to few candidate genes. Here, we describe the detailed step-by-step guide for performing MetaQTL specific regional association analysis to identify important genomic regions and underlying potential major effect genes governing traits of agronomic importance in cereals.

Key words Association mapping, Fine mapping, Linkage mapping, MetaQTL, QTL

1 Introduction

Developing crop varieties with higher yield, superior nutritional composition and with ability to withstand a diverse range of biotic and abiotic stresses are pivotal for the sustenance of rapidly growing human population [1, 2]. Conventional plant breeding has provided a means to develop such superior crop varieties for decades. However, with high cost and time associated with conventional plant breeding, it is unlikely to fulfill the unprecedented need for superior crop varieties especially in context of rapidly changing climatic scenarios [3]. Understanding the genetic and molecular basis of important agronomic and quality traits provides opportunity to employ marker-assisted selection to speed up the breeding program in a cost-effective

Luis M. Vaschetto (ed.), *Cereal Genomics: Methods and Protocols*, Methods in Molecular Biology, vol. 2072, https://doi.org/10.1007/978-1-4939-9865-4_3,

manner [4, 5]. Linkage mapping using biparental or multiparental mapping populations is one of the most utilized and powerful methods to identify genomic regions regulating important agronomic traits in cereals [6]. So far, hundreds of QTLs (quantitative trait loci) have been identified in almost all major economically important food crops for several agronomically useful traits [7–9]. However, these identified QTLs usually span long genomic intervals, harboring hundreds/thousands of genes. Till now, very few of these QTLs could be narrowed-down to single major effect gene(s), as fine mapping requires genotyping of hundreds to thousands of mapping individuals and therefore requires significant cost, time and labor. This has seriously limited the utility of these QTLs for marker-assisted genetic improvement of cereals [10]. Apart from linkage mapping, a NGS based version of bulk segregant analysis, popularly known as QTLseq approach, has been recently developed. QTLseq due to its rapid nature is becoming increasingly popular and has been successfully employed for major QTL identification in a wide range of crops in last few years [11–13]. However, similar to conventional linkage mapping, the QTLs identified using QTLseq also need to be fine mapped to one or few candidate gene(s) for their efficient utilization in marker-assisted breeding for crop genetic enhancement.

Genome-wide association study (GWAS) is another popular approach for identification of genomic loci (QTLs) which on contrary to linkage mapping provides much higher resolution (smaller QTL intervals with comparatively fewer candidate genes) and requires lesser time (as it eliminates the need for generation of experimental mapping populations) [14–16]. Despite these advantages, QTLs identified with GWAS are seldom suitable for marker-assisted selection, except in cases where the underlying causal gene/polymorphism has been successfully identified. Apart from this, association mapping also suffers from very high level of false positives majorly due to presence of population structure and cryptic relatedness among accessions (association panel) used for association mapping [17, 18]. These disadvantages associated with association mapping limit its practical applicability in crop improvement endeavors.

Despite their respective limitations, combined use of these two methods provides unique opportunity to bring together advantages associated with both linkage mapping (low false positives and high power of QTL detection) and association mapping (high mapping resolution). This can enable us to identify and narrow down authentic genomic regions governing important agronomic traits and which further can be narrowed down to single major effect gene using diverse strategies. Considering this, joint linkage-

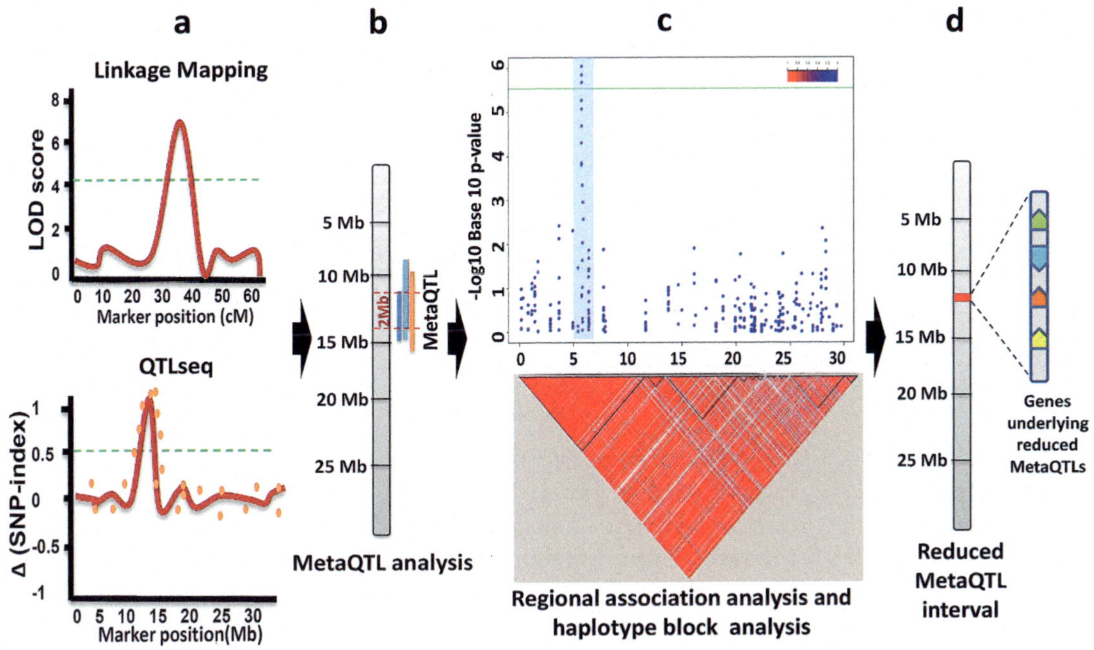

Fig. 1 Schematic diagram depicting metaQTL region specific association analysis to narrow down long intervals of known QTLs to short QTL genomic region(s) with few candidate genes governing traits of agronomic importance. (**a**) QTLs detected with linkage mapping and QTLseq analysis, (**b**) meta-analysis of known QTLs to detect metaQTL, (**c**) regional association analysis and haplotype block analysis of metaQTL, (**d**) reduced metaQTL interval harboring only few candidate genes

association analysis and/or QTL region-specific association analysis have been developed and have been successfully employed in different cereals [19, 20]. Recently, we developed a strategy called "meta-QTL region specific regional association analysis" which combines meta-QTL analysis and regional association mapping [21] (Fig. 1). We also demonstrated its utility to efficiently scale down previously known grain size/weight QTL intervals to one or few genes in rice [21]. In contrary to original identified QTLs, new reduced QTL intervals or genes can be efficiently utilized in marker-assisted selection for rice crop improvement. Here, we provide in depth details of approach called "meta-QTL region specific regional association analysis" which we have used to identify and narrow-down metaQTL regions associated with grain size in rice. This optimized strategy can be used for further narrowing-down the long interval QTLs into either short QTL genomic regions or specific gene(s) governing yield and a/biotic stress tolerance traits, thus has broader practical applications for marker-aided genetic improvement of cereals.

2 Materials

1. System requirement: Unix/Linux operating system with Java installed.
2. R packages: MASS, multtest, genetics, compiler, scatterplot3d, gplots, and GAPIT.
3. Software packages: BioMercator, HapQTL, and Haploview.
4. Dataset: QTLs and SNPs for multiple accessions of cereals.

3 Methods

3.1 Selection of Authentic QTLs for Meta-analysis

Selection of suitable QTLs for meta-analysis is one of the first and most crucial steps for success of meta-QTL specific regional association analysis. Generally, all the QTLs reported for the trait of interest are identified through extensive literature survey. However, it is important to have few considerations in mind while selecting QTLs for meta-analysis (*see* **Notes 1–7**). The general steps for selection of QTLs are as follows:

1. Conduct comprehensive literature survey and compile all the available QTL information for the trait of interest (yield, quality, biotic and abiotic stress, etc.). This can be best performed by retrieving relevant publications using NCBI's PubMed search engine (https://www.ncbi.nlm.nih.gov/pubmed/) and subsequently compiling QTL information such as flanking markers, PVE (phenotypic variation explained) and LOD (logarithm of odds) score in tabular format (*see* **Note 1**). QTL information databases like Gramene (http://archive.gramene.org/qtl/), QTARO (http://qtaro.abr.affrc.go.jp/), and CSGRqtl (http://helos.pgml.uga.edu/qtl/) can also be utilized to rapidly retrieve QTL information of multiple cereals.
2. Extract physical coordinates of all the QTLs by performing BLAST using the primer sequences of flanking markers (both left and right markers associated with QTLs) against the latest version of reference genome of cereal available.
3. Filter compiled QTL information based on variety of criteria including PVE, LOD, and length of QTL interval and eliminate all QTLs which do not fulfill the set criteria (*see* **Notes 2–6**).
4. Eliminate any QTL for which either physical coordinates cannot be determined with confidence or they have an interval longer than the threshold (*see* **Note 7**).

3.2 Meta-analysis of Selected QTLs

After selection of suitable QTLs, the next step is to perform meta-analysis to identify genomic regions repeatedly identified in independent QTL mapping studies. This provides us with authentic genomic regions associated with the trait of interest. All the selected QTLs in the previous step can be further subjected to meta-analysis to identify meta-QTLs associated with the trait of interest. Meta-analysis can be performed using BioMercator (Version 4.1) software with the following simple steps:

1. Download and open Biomercator software using the following link. (http://moulon.inra.fr/index.php/fr/component/docman/cat_view/21-logiciels/101-abi-project-and-software/104-biomercator-v21).
2. Create new project: File → Genetic data loading → new project → project name.
3. Choose newly created project: File → Genetic data loading → project name.
4. Load map file: File → Genetic data loading → project name → Load Map.txt file (*see* **Note 8**).
5. Load QTL file: File → Genetic data loading → project name → Load QTL.txt file (refer Notes for creation of QTL file).
6. Perform MetaQTL analysis (multi steps process): (a) Analyses → QTL Meta analyses → MetaQTLMeta analysis ½, (b) Analyses → QTL Meta analyses → MetaQTLMeta analysis ½ → select desired map → select desired LG → select desired Chromosome → regroup the traits into a single meta trait → launch analysis, and (c) Analyses → QTL Meta analyses → MetaQTLMeta analysis 2/2 → select project → select map → select chromosome → select linkage group → select meta-analysis name given in previous step → select meta trait name given in previous step → provide number of metaQTLs for the model as suggested by model.txt.
7. Extract metaQTL intervals from _tables.txt file.
8. Screen all the identified metaQTLs for the presence of known genes related to the trait of interest within their confidence interval and eliminate all metaQTLs harboring any known gene regulating the trait of interest. The remaining metaQTL intervals can be used for further analysis.

3.3 Regional Association of metaQTL Regions

After identification of metaQTL regions, the regional association analysis of each identified metaQTLs is conducted to further narrow down the metaQTL intervals. To conduct association analysis, we require sequence variation data for hundreds of diverse natural germplasm accessions with sufficient phenotypic diversity for trait of our interest. For many crops including rice, maize, and soybean, the genome-wide sequence variation data has been made publicly available in the interest of global research community and can be directly retrieved and used for GWAS. However, for crops for which sequence variation data is not available yet, the metaQTL regions can be genotyped in a set of diverse natural germplasm accessions using various targeted genotyping assays. Regional association analysis can be performed using commonly adopted single SNP based methods widely used for GWAS in cereals. However, in addition to single SNP based methods, haplotype-based methods can also be used to perform association analysis, as they can account for locus as well as allelic heterogeneity and therefore can detect many novel associations which are usually missed with single SNP-based association analysis.

3.3.1 Regional Association Analysis

1. Retrieve sequence variation data for diverse natural accessions phenotyped for the trait of interest (if publicly available) or perform genotyping of known markers (SNPs/INDELs/SSRs) which are present within metaQTL regions, in diverse natural accessions phenotyped for the trait of interest.
2. Filter the genotype data for missingness and eliminate all accessions or site showing high degree of missingness.
3. Perform regional association analysis by using filtered genotype data using appropriate method.

3.3.2 Association Analysis Using GAPIT

Single SNP based association can be performed using Compressed Mixed Linear Model (Compressed MLM) implemented in Genome Wide Association and Prediction Tool (GAPIT). Running GAPIT requires R to be installed on your computer. The following code can be used to install and run GAPIT.

```
#Install packages (Required for new installation only)
#---------------------------------------------------------------
source("http://www.bioconductor.org/biocLite.R")
biocLite("multtest")
install.packages("gplots")
install.packages("scatterplot3d")#The downloaded link at:
http://cran.r-project.org/package=scatterplot3d
Step 0: Import library and GAPIT functions (run this section
every time you run GAPIT)
#####################################################
library('MASS')
library(multtest)
library(gplots)
library(compiler)
library("scatterplot3d")
source("http://www.zzlab.net/GAPIT/emma.txt")
source("http://www.zzlab.net/GAPIT/gapit_functions.txt")
```

```
#source("/Users/Zhiwu/Dropbox/Current/revolutionr/gapit/gapit_
functions.txt")
######################################################
# Save genotype and phenotype files (refer GAPIT tutorial data)
in myDATA directory
setwd("/myDATA")
#Running Compressed MLM [22]
#-------------------------------------------------------------
#Step 1: Import genotype and phenotype files
myY<- read.table("phenotypes.txt", head = TRUE)
myG<- read.delim("genotype.hmp.txt", head = FALSE)
#Step 2: Run GAPIT
myDATA<- GAPIT(
Y=myY,
G=myG,
PCA.total=3)
```

- The results can be accessed from .GWAS.Results.csv file generated in working directory.
- All SNPs with FDR adjusted P-values less than 0.05 are considered to be significantly associated with the studied trait.
- Conduct haplotype block analysis using haploview software (https://www.broadinstitute.org/haploview/tutorial) to determine genomic region (QTL interval) represented by the most significant SNP.

3.3.3 Haplotype-Based Association Analysis

Although single SNP-based association analyses are useful for identifying common variants, they often fail to detect rare variants associated with the trait of interest. Haplotype-based or multi-SNP-based association methods can be used to overcome above mentioned bottleneck as they can account for allelic heterogeneity (different mutations within same gene causing same phenotype) as well as locus heterogeneity (mutations in two different genes causing same phenotype) [23–25]. HapQTL software implements a novel hidden Markov model method for calculating local haplotype sharing defined as the probability of two diploid individuals descending from the same ancestral haplotypes or genetic similarity between individuals. Further it tests association between genetic similarity and phenotype to detect association between the local haplotype and a phenotype. Association analysis in hapQTL can be performed using the following steps:

1. Create and save all input files (genotype file, SNP position file, phenotype file, and covariate file) as per format detailed in the hapQTL manual.
2. The following command line can then be used to run hapQTL

/hapqtl -g genotype_file.txt -pos SNP_position.txt -w 30 -C 3 -c 10 -mg 200 -exclude-maf 0.01 --exclude-nopos -sem 1 –pk1 -o pref -FILE pheno.txt -DOC cov.txt (details about the EM parameters can be referred from the hapQTL manual).

3. The above command will generate four output files: prefix.log (contains details of parameters used during the run), prefix.snpinfo.txt (contains information on SNP position and allele frequencies), prefix.bf.txt (Bayes factor file contains Bayes factor and p-value for haplotype and single SNP), and prefix.pk1.txt file (LHS matrix file).
4. The prefix.pk1.txt file can be used to calculate mLHS using the following R code.

```
##################################################################
ns=412; ## note this is total number of snps in the region.
s=226; ##note this is the top-hit (can be obtained from prefix.
bf.txt) .
bb=rep(0,ns);
for (try in c(1:10)) {
fn=paste('output/region.ra.chr11-120-em1-C2-c10-w30-s0-mg100-
miss5pc-3snp.try',try,'.pk1.txt',sep='');
xx=scan(fn);
dim(xx)=c(10,ns,4476) ##note here 4476 is number of individu-
als, and 10 is number of lower-layer clusters.
yy=xx[,s,]
aa=rep(0,ns);for(i in c(1:ns)){zz=yy*xx[,i,];vz=apply(zz,2,sum);
aa[i]=mean(vz);}
bb=bb+aa;}
bb=bb/10;
write(bb, "lhs.txt",1);

#########################################################
##lhs.txt contains the pair-wise lhs between the top-hit mark-
ers and the rest.
##hap.txt contains the most-significant ancestral haplotypes,
which can be obtained from pref.em.txt
##snpinfo.txt contains snp position info.
lhs=scan('lhs.txt');
hap=read.table('hap.txt');
snp=read.table("output/region.ra.chr11-120-em1-C2-c10-w30-s0-
mg100-miss5pc-3snp.try9.snpinfo.txt", 1);
vgray=colors()[grep("gray",colors())][10:109];

par(mfrow=c(1,1));
par(mar=c(4,4,2,2));
pdf('ra-chr11-120.pdf', width=12, height=8);
ss=c(1:length(lhs))[lhs>0.7]
```

```
plot(snp$pos/1e6, lhs/4, type="b", lwd=1.5, cex=0.6,pch="x",
xlim=c(119.9, 120.5), ylim=c(0,0.6), col="darkgray",
xlab="Chromosome 11 (Mb)", ylab="mLHS", cex.lab=1.2)
mp=which(lhs==max(lhs));
points(snp$pos[mp]/1e6, lhs[mp]/4, cex=1, pch="+",col=1);
points(snp$pos[ss]/1e6, rep(c(0.10), length(ss)), col=vgray[(1-
hap[,1])*90+1], cex=1, pch=19)
points(snp$pos[ss]/1e6, rep(c(0.12), length(ss)), col=vgray[(1-
hap[,2])*90+1], cex=1, pch=19)
segments(120.036, 0.05, 120.362, 0.05, lwd=2)
text(120.21, 0.035, "GRIK4", cex=1.2)
text(120.4, 0.5, "RA", cex=1.2);
dev.off()
###The haplotype block detected by calculating mLHS around the
most significant SNP can be considered as QTL.
```

3.4 Candidate Gene Identification Within QTL Interval

Once the QTL region is narrowed down after regional association analysis, the underlying major effect genes can be identified by applying a diverse range of strategies some of which are as follows:

1. The putative functions of the candidate genes within reduced metaQTL interval can be identified with similarity search against protein sequences of closely related species/model plants.
2. Search for any major effect variants present within reduced metaQTL interval as gene harboring such variants are likely to be the major effect genes.
3. In silico expression analysis can also be performed to check expression profiles of all genes within reduced metaQTL intervals. This can provide important information about the probable function of the genes.

4 Notes

1. The study reporting QTLs must have a published record in peer-reviewed journal of national or international repute.
2. The study should be based on sequence-based markers and the primer sequences of markers used for the QTL identification must be available as a part of publication or in some other publicly available open access database.
3. The QTL mapping results must be based on multiyear/multilocation field phenotyping data for the trait of interest.
4. Based on the goal of the study, suitable PVE (phenotypic variation explained) threshold can be set. For example, for identifi-

cation/narrow down of major QTLs, PVE threshold of ≥10% can be used.

5. Similar to PVE, suitable LOD (logarithm of odds) threshold can also be set to select QTLs.
6. The primer sequences for left and right flanking marker associated with QTLs should map uniquely on same chromosome of reference genome so as to extract physical coordinates for the QTL interval.
7. The length of QTLs as derived from their physical coordinates must not exceed 1/5th of average chromosome length for a concerned crop.
8. The map file for metaQTL analysis can be created with evenly distributed dummy markers. The format for map.txt file can be referred from the BioMercator's user manual.

Acknowledgments

The work in lab is supported by grants from DBT and SERB, Government of India.

References

1. Tester M, Langridge P (2010) Breeding technologies to increase cropping production in a changing world. Science 327:818–822
2. Alexandratos N, Bruinsma J (2012) World agriculture towards 2030/2050: the 2012 revision. FAO: ESA Working Paper No. 12-03.
3. Moose SP, Mumm RH (2008) Molecular plant breeding as the foundation for 21st century crop improvement. Plant Physiol 147:969–977
4. Dar MH, de Janvry A, Emerick K et al (2013) Flood-tolerant rice reduces yield variability and raises expected yield, differentially benefitting socially disadvantaged groups. Sci Rep 3:3315
5. Ellur RK, Khanna A, Yadav A et al (2016) Improvement of Basmati rice varieties for resistance to blast and bacterial blight diseases using marker assisted backcross breeding. Plant Sci 242:330–341
6. Doerge RW (2002) Mapping and analysis of quantitative trait loci in experimental populations. Nat Rev Genet 3:43–52
7. Fan C, Xing Y, Mao H et al (2006) *GS3*, a major QTL for grain length and weight and minor QTL for grain width and thickness in rice, encodes a putative transmembrane protein. Theor Appl Genet 112:1164–1171
8. Ronald PC, Albano B, Tabien R et al (1992) Genetic and physical analysis of the rice bacterial blight disease resistance locus, *Xa21*. Mol Gen Genet 236:113–120
9. Gamuyao R, Chin JH, Pariasca-Tanaka J et al (2012) The protein kinase Pstol1 from traditional rice confers tolerance of phosphorus deficiency. Nature 488:535–539
10. Kumar J, Gupta DS, Gupta S et al (2017) Quantitative trait loci from identification to exploitation for crop improvement. Plant Cell Rep 36:1187–1213
11. Takagi H, Abe A, Yoshida K et al (2013) QTL-seq: rapid mapping of quantitative trait loci in rice by whole genome resequencing of DNA from two bulked populations. Plant J 74:174–183
12. Shu J, Liu Y, Zhang L et al (2018) QTL-seq for rapid identification of candidate genes for flowering time in broccoli × cabbage. Theor Appl Genet 131:917–928
13. Srivastava R, Upadhyaya HD, Kumar R et al (2017) A multiple QTL-Seq strategy delineates potential genomic loci governing flowering time in chickpea. Front Plant Sci 8:1105
14. Huang X, Han B (2014) Natural variations and genome-wide association studies in cereals. Annu Rev Plant Biol 65:531–551

15. Xiao Y, Liu H, Wu L et al (2017) Genome-wide association studies in maize: praise and stargaze. Mol Plant 10:359–374
16. Zhou Z, Jiang Y, Wang Z et al (2015) Resequencing 302 wild and cultivated accessions identifies genes related to domestication and improvement in soybean. Nat Biotechnol 33:408–414
17. Teo YY (2008) Common statistical issues in genome-wide association studies: a review on power, data quality control, genotype calling and population structure. Curr Opin Lipidol 19:133–143
18. Korte A, Farlow A (2013) The advantages and limitations of trait analysis with GWAS: a review. Plant Methods 9:29
19. Lu Y, Zhang S, Shah T et al (2010) Joint linkage-linkage disequilibrium mapping is a powerful approach to detecting quantitative trait loci underlying drought tolerance in maize. Proc Natl Acad Sci U S A 107:19585–19590
20. Wu X, Li Y, Shi Y et al (2016) Joint-linkage mapping and GWAS reveal extensive genetic loci that regulate male inflorescence size in maize. Plant Biotechnol J 14:1551–1562
21. Daware AV, Srivastava R, Singh AK et al (2017) Regional association analysis of MetaQTLs delineates candidate grain size genes in rice. Front Plant Sci 8:807
22. Zhang Z, Ersoz E, Lai CQ et al (2010) Mixed linear model approach adapted for genome-wide association studies. Nat Genet 42:355–360
23. Pritchard JK (2001) Are rare variants responsible for susceptibility to complex diseases? Am J Hum Genet 69:124–137
24. Zöllner S, Pritchard JK (2005) Coalescent-based association mapping and fine mapping of complex trait loci. Genetics 169:1071–1092
25. Guan Y, Stephens M (2011) Bayesian variable selection regression for genome-wide association studies, and other large-scale problems. Ann Appl Stat 5:1780–1815

Chapter 4

Sequencing and Assembling Genomes and Chromosomes of Cereal Crops

Marcelo Helguera

Abstract

Next generation sequencing (NGS) and assembly are key pieces in the success of cereal crops genomes sequencing. One interesting strategy for reducing the complexity in large genomes, the case of several cereal crops, is the sequencing of individual chromosomes. This has been done with success by flow cytometric chromosome sorting followed by sequencing using available next generation (high throughput) sequencing platforms. In this chapter, methodologies for sequencing and assembly of flow sorted chromosomes and whole genomes in cereal crops, with special emphasis on the case of the International Wheat Genome Sequencing Consortium (IWGSC), are reviewed.

Key words Cereal crops, Chromosome and genome sequencing, Assembling

1 Introduction

Next generation sequencing (NGS) is a key piece in the success of large plant genomes sequencing, but not the only one. With an haploid content of 16 gigabases (Gb) [1] the bread wheat (*Triticum aestivum* L, 2N = 6× = 42 subgenomes AABBDD) genome full annotation was conceived by sequencing of individual chromosomes as strategy to reduce genome complexity taking advantage of the flow cytometric chromosome sorting [2]. Flow cytometry is designed to analyze microscopic particles during a passage in a narrow stream of liquid based on optical properties (light scatter, fluorescence). Flow cytometry has been used successfully to analyze and sort metaphase chromosome from nuclear genomes of human, animals and plants [3]. In the case of bread wheat obtaining a full annotation of the gigant genome involved the organization of an international team, the International Wheat Genome Sequencing Consortium (IWGSC), and the coordinated work for more than 12 years of research groups from different part of the world. The first gold-standard wheat genome sequence release publication included the contribution of 73 research groups covering

Luis M. Vaschetto (ed.), *Cereal Genomics: Methods and Protocols*, Methods in Molecular Biology, vol. 2072, https://doi.org/10.1007/978-1-4939-9865-4_4,

whole-genome sequencing and assembly; gene and transposable elements annotation and curation; transcriptome analyses; microRNA and tRNA annotation; genetic maps and mapping; chromosome sorting and BAC libraries; individual chromosome BAC sequencing and assemblies, among other analyses [4]. In this study, with an estimated coverage of 94%, 107,891 high-confidence gene models were annotated resolving issues like the intergenic space and presenting the genome in the correct physical order [4]. The freely available wheat genome reference sequence IWGSC RefSeq v1.0 is a disruptive tool for wheat breeding in terms of accelerating QTL mapping and gene discovery through the physical anchoring of genetic maps and a first survey of annotated candidate genes. Examples of QTL mapping/validation studies anchoring genetic data to the IWGSC RefSeq v1.0, and eventually a preliminar list of candidate genes, can be found for different traits including, disease resistance [5–10], adaptation and yield components [11–13], and food and nutrition quality [14–16], among others.

Moreover, single chromosome survey sequencing is a powerful tool for gene discovery in particular cases of unsequenced specific cultivars [17, 18], presence of wheat-wild relatives chromosome translocations, and for plant/crop models where a reference complete genome sequence is still unavailable, like rye (*Secale cereale*) and oat (*Avena sativa*) within the *Poaceae* Family.

In this chapter, methodologies for sequencing and assembly of single chromosomes and whole genomes in cereal crops are reviewed.

2 Methodologies for Chromosome Sorting and DNA Amplification

The first step for NGS is to increase DNA content of target chromosomes, which can be achieved successfully through chromosome sorting by flow cytometry and DNA amplification by multiple displacement amplification (MDA) [19–24]. In this approach a relatively low number of chromosomes (between 10,000 and 160,000 chromosome or chromosome arms) is isolated by flow cytometry to be later amplified by MDA. A typical MDA reaction yields 3–6 micrograms of DNA with average length between 5 and 30 kb [25], which is suitable for applications such as RFLP, SNPs genotyping, physical mapping [25–27] and more recently, single chromosome survey sequencing [19–24]. Additional methods for chromosome sorting have been described including microdissection [28] and isolation of mitotic chromosomes in suspension by capture on magnetic beads following hybridization with a labeled chromosome-specific probe [29]. However, in comparison with flow cytometry, previously mentioned methods undergo some limitations, as being highly labor intensive or suffering low purity in isolated fractions [3, 30].

3 Methodologies for NGS

Two of the most widely used NGS platforms are 454 Roche and Illumina. Illumina delivers short reads of 36 base pairs (bp) in size at low cost, estimated in $29/Gb based on the Hi Seq 2500 HT v4 platform [31, 32]. 454 Roche platform delivers longer reads of up to 1 kilobase (kb) in size, but at a much higher cost per base, estimated in $8000/Gb based on GS FLX+ platform [33]. Additional short read NGS platforms available at the market are Ion Torrent (Thermo Fisher) delivering 200 bp reads at the cost of $100/Gb based on Ion Proton platform [34, 35] and SOLiD (Thermo Fisher) delivering 2× 50 bp reads at the competitive cost of $17/Gb based on SOLiD 5500×l Wildfire platform [36]. There is also a second group of NGS platforms delivering much longer reads and includes PacBio and Oxford Nanopore, among others. PacBio (Pacific Biosciences), delivers reads with an average length above 10 kb at the cost of $85/Gb based on Sequel platform [37, 38], and the Oxford Nanopore sequencer delivers very long reads of up to 2 megabases (Mb) in size at the competitive cost of $15.75/Gb based on MinION Mk 1B sequencer [39]. The case of MinION sequencer is interesting because with an equipment weight of 87 g, and a power requirement of one USB3 port it can be used to process samples in the field. Table 1 describes technical specifications of NGS platforms referred before.

In general terms, long read sequencing is seen as a powerful tool to deliver long-range genomic information and, in this way, spam repetitive regions which are difficult to rebuild using short reads data. Besides these advantages, long read sequencing adoption was initially hampered by low throughput and relatively high error rates estimated in 5–10% for Oxford technologies and 14% for PacBio vs. 0.1–1% for leading short reads platforms [38, 39]. Recent advances in long reads associated technologies have reduced the high error rate issue, including hybrid assembly methods using Illumina reads to correct error like NanoCorr [40], Pilon [41] and Nanopolish [42] among others. Another interesting tool to improve whole genome assemblies is optical mapping, a light microscope-based technique that uses restriction site images to produce high definition fingerprints of DNA sequences [43]. Optical mapping offers the ability to assay long DNA molecules (20 kb to 3 Mb depending to the technology) leading to the identification of frequent missassemblies in large genomes such as false joins, inversions, and translocations by the comparison of consensus optical maps with the assembly [44]. Optical mapping has been used in the building cereal crops genome assemblies including *Sorghum bicolor* [45], *Zea mays* [46] and *Hordeum vulgare* [47] among others.

Table 1
Comparison of NGS platforms specifications[a]

	Illumina Hi Seq 2500 HT v4	Roche 454 GS FLX	Ion Proton PI	SOLiD 5500 Wildfire	PacBio Sequel	MinION Mk 1B[b]
Total output	1 Tb	700 Mb	10 Gb	120 Gb	10 Gb	30 Gb
Run time	6 days	23 h	2–4 h	10 days	4 h	2 days
Read length	2 × 125 bp	Up to 1 kb	Up to 200 bp	2 × 50 bp	10–15 kb	Up to 2 Mb
# of single reads	4000 M	1 M	Up to 82 M	1200 M	0.5 M	512
Instrument price[c]	$740,000	~$500,000	$150,000	$420,000	$350,000	–
Run price[c]	~$2900	~$6000	$1000	$2050	$850	$475
$/Gb[c]	$29	~$8000	$100	$17	$85	$15.75

[a]Information based on http://allseq.com/kb-category/sequencing-platforms/, with modifications
[b]Instrument price, Run price and $/Gb are in US$ dollars, April 22 2019
[c]Information based on https://nanoporetech.com/products/comparison

4 Chromosome and Genome Sequencing and Assembly, Cases Studies in Cereal Crops

The wheat whole genome survey sequencing (incomplete sequence coverage) based on single chromosome arms isolation, sequencing and assembly was first published in 2014 [48]. In this approach single chromosome arms were purified by flow-cytometric sorting as previously described (see **item 2** in this chapter) and sequenced with an Illumina HiSeq 2000 platform to a depth of 30 to 241× using paired-end (PE) sequence libraries with 500 bp target size. In this case PE sequence reads were assembled with the short-reads assembly tool ABySS [49]. ABySS as other de Bruijn graph assemblers are much more efficient in assemblying short reads (less than 50 bp) typical of Illumina, Ion Torrent and SOLid Platforms, than reads produced by Roche 454 technology (>100 bp) and long reads (more than 10 kb) generated by PacBio and Oxford Nanopore platforms [50]. Additional assemblers using the de Bruijn graph algorithm are Velvet [51] and SOAPdenovo [52], among others. Roche 454 is a particular case, delivering slightly larger reads (100–500 bp) than Illumina, the company developed a non- de Bruijn custom-made assembler named Newbler [53].

Taking as example 4D chromosome, cumulative data of contig assemblies obtained with Illumina [48] showed slightly better numbers than Roche 454 technology [19]: contig sizes of 142.1 and 347.6 Mb for chromosome arms 4DS and 4DL with N50 values of 3278 and 1013 bp, respectively (N50 is the smallest contig size required to cover 50% of the assembly) obtained using the Illumina platform were higher than contig sizes of 103 and 120 Mb for chromosome arms 4DS and 4DL (N50 values of 1132 and 807 bp, respectively) obtained with 454 technology. A possible explanation for the better performance of Illumina vs. 454 technology assemblies can be traced to differences in the depth of sequencing coverage, with values of 30 to 241× in the case of Illumina vs. 7.2× and 4.1× for short and long 4D chromosome arms in the case of 454 technology. Higher sequencing redundancy values in 454 sequencing will probably provide better contig coverage and N50 values. Examples of single chromosome survey sequencing based on 454 technology are chromosome arms 5AS 2.44× and N50 = 803 bp, 5AL 2.35× and N50 = 798 bp [20], 6BS 11.9× and N50 = 1107 bp, 6BL 11.1× and N50 = 2675 bp [22] among others, ensambled with different newbler versions (see Table 2). Examples based on Illumina platform are chromosome arms 7BS 21× and N50 = 472 [24], 7DS 34× and N50 = 1159 bp [23] and *T. dicoccoides* chromosomes, with 39–58× coverage and N50 = 1092-4938 bp [54] ensambled with de Bruijn based assemblers ABySS and Velvet (Table 2). Interestingly, when considering gene content, the information obtained with Roche 454 data [19] and Illumina [48] showed 21–28% of 454-specific loci and 20–35% Illumina-specific loci. As previously expressed [19] these data suggest a complementary coverage, at least in syntenic gene content, between Roche 454 and Illumina technology.

Recent progress in NGS considering items like read size, assembling algorithms, gap filling, scaffolding, error rate polishing, and sequencing costs, have radically improved and facilitated the accurate sequencing of complete genomes including cereal crops. This is the case of PacBio and Oxford Nanopore, single-molecule sequencing technologies that produce reads longer than 10 kb in average demanding a different set of assemblers including Canu [55], Falcon [56], ALLPATHS-LG [57] and DBG2OLC [58] among others, in many cases using an hybrid approach combining long and short reads for sequence polishing in the assembly of genome sequences.

The impact of previously described advances in whole genome assembling can be pictured in the case of rice, the first Poaceae member with the nuclear genome sequenced [59]. In this fundamental work the rice nuclear genome was sequenced with Sanger technology (previous NGS) using the whole genome shotgun sequencing approach. With a 4× sequence coverage the obtained assembly covered 362 Mb with N50 = 11.76 kb. Fifteen years later, the rice genome sequencing shows major improvements in terms of base error rates (<0.0017%) delivering a near complete genome of *Oryza sativa* var. *indica* [60]. In this example, a hybrid assembly strategy was performed including PacBio (118×) and

Table 2
Comparisons considering assemblies statistics for *T. aestivum* single chromosome arms

Target	DNA size	Sequencing platform	Sequence coverage	Assembly tool	Assembly size	N50	Reference
4DS	232 Mb	454	7.2×	Newbler v. 2.6	103 Mb	1132 bp	[19]
4DL	417 Mb	454	4.1×	Newbler v. 2.6	120 Mb	807 bp	[19]
4DS	232 Mb	Illumina	>30×	ABySS	142.1 Mb	3278 bp	[48]
4DL	417 Mb	Illumina	>30×	ABySS	347.6 Mb	1013 bp	[48]
5AS	295 Mb	454	2.44×	Newbler v. 2.3	61 Mb	803 bp	[20]
5AL	532 Mb	454	2.35×	Newbler v. 2.3	116 Mb	798 bp	[20]
6BS	415 Mb	454	11.9×	Newbler v. 2.7	234.7 Mb	1107 bp	[22]
6BL	498 Mb	454	11.1×	Newbler v. 2.7	273.2 Mb	2675 bp	[22]
7BS	360 Mb	Illumina	21×	Velvet v. 1.0.09	176.1 Mb	472 bp	[24]
7DS	381 Mb	Illumina	34×	Velvet v. 1.0.09	153.6 Mb	1159 bp	[23]
T. diccocoides 4A	864 Mb	Illumina	41×	ABySS	388.6 Mb	1114 bp	[54]

Illumina (>198×) reads delivered from (1) whole genome libraries, (2) genotyping by sequencing (GBS) fosmid libraries and (3) GBS of a mapping population combined with optical mapping data. PacBio long reads were polished and assembled with PBcR. Illumina short reads were assembled with SOAPdenovo in the case of whole genome libraries. GBS data from fosmid libraries and mapping populations was aligned to the PBcR assembly using BWA-aln and BWA-mem in the building of a supercontigs that were covalidated with optical mapping combining PBcR assembly with Bionano maps (Table 3). Additional examples of cereal crop genomes sequencing strategies including sorghum, maize, barley and wheat assembling can be found in Table 3.

Also it is worth mentioning the assembler DeNovoMAGIC2.0 [61] which works with short- reads (Illumina) to get accurate sequence of long inserts for large genome assembly and characterisation. DeNovoMAGIC2.0 was used successfully to get the gold-standard wheat genome sequence [4]. The ordered and annotated assembly IWGSC RefSeq v1.0 included 14.5 Gb leaded

Table 3
Comparisons considering assemblies statistics for whole genomes cereal crops

Target	DNA size	Sequencing platform	Sequencing strategy	Sequence coverage	Assembly tools	Assembly size	N50	Reference
Oryza sativa	430 Mb	MegaBACE 1000 (Sanger)	WGSS[a] plasmid libraries	4×	RePS	362 Mb	11.76 kb	[59]
O. sativa	390 Mb	Illumina and PacBio	WGSS illumina and PacBio libraries + GBS[b] fosmid libraries + GBS mapping population + optical mapping	198×	PBcR + SOAPdenovo + optical mapping	382.6 Mb	2.48 Mb	[60]
Sorghum bicolor	730 Mb	Sanger	WGSS and BAC by BAC approach	8.5×	Arachne2	625.7 Mb	64.3 Mb	[62, 63]
S. bicolor	732 Mb	ONT and Illumina	WGSS + optical mapping	232.7×	Canu+SMARTdenovo + Pilon + optical mapping	661.1 Mb	33.28 Mb	[45]
Zea mays	2.3 Gb	Sanger	BAC by BAC approach	4–6×	N.A.	2.05 Gb	76 kb	[64]
Z. mays	2.3 Gb	PacBio and Illumina	WGSS + optical mapping	165×	PBcR –MHAP + Falcon + Pilon + optical mapping	2.1 Gb	9.56 Mb	[46]
Hordeum vulgare	5.1 Gb	Sanger (BAC ends), 454 and Illumina	BAC by BAC approach and WGSS	14× (BAC by BAC) >50× WGSS	MIRA (454); VELVET, CLC (Illumina)	4.98 Gb	904 kb	[65]
H. vulgare	5.1 Gb	454 and Illumina	BAC by BAC approach	1000×	MIRA (454); ABySS, SOAPdenovo, SOAPaligner, SSPACE, CLC (Illumina) + optical mapping	4.54 Gb	1.9 Mb	[47]
T. aestivum	17 Gb	Illumina	Shotgun sequencing chromosome arm scale	>30–241×	ABySS	10.2 Gb	1.7–8.9 kb	[48]

(continued)

Table 3
(continued)

Target	DNA size	Sequencing platform	Sequencing strategy	Sequence coverage	Assembly tools	Assembly size	N50	Reference
T. aestivum	17 Gb	Illumina and PacBio	Shotgun sequencing chromosome arm scale	101×	w2rap-contigger and SOAPdenovo	13.4 Gb	88.8 kb	[66]
T. aestivum	17 Gb	Illumina	BAC by BAC combined with shotgun sequencing chromosome arm scale	234×	DeNovoMAGIC2.0	14.5 Gb	22.8 Mb	[4]

[a]Whole genome shotgun sequencing
[b]Genotyping by sequencing

an average N50 values of 52 kb (contigs), 7 Mb (scaffolds) and 22.8 Mb (superscaffolds) with an estimated coverage of 94% of the genome (Table 3).

5 Main Conclusions

1. Single chromosome sequencing is a highly efficient strategy to reduce complexity of large genomes like bread wheat. This approach is a powerful tool for gene discovery in unsequenced specific cultivars, presence of cereal-wild relatives chromosome translocations, and for plant/crop models where a reference complete genome sequence is still unavailable.
2. Recent progress in NGS in terms of reads size, assembling algorithms, sequence polishing (hybrid assemblers), integration of physical information at chromosome level (optical mapping) and lower sequencing costs have radically improved the accurate sequencing of complete genomes in cereal crops. The most relevant achievement of this technological revolution in cereal crops is the recent release of the gold-standard wheat genome sequence IWGSC RefSeq v1.0.

References

1. Arumuganathan K, Earle ED (1991) Nuclear DNA content of some important plant species. Plant Mol Biol Rep 9:208–218
2. Vrána J, Kubaláková M, Simková H et al (2000) Flow sorting of mitotic chromosomes in common wheat (*Triticum aestivum* L.). Genetics 156:2033–2041
3. Doležel J, Vrána J, Šafář J et al (2012) Chromosomes in the flow to simplify genome analysis. Funct Integr Genomics 12:397–416
4. International Wheat Genome Sequencing Consortium (IWGSC), IWGSC RefSeq principal investigators, Appels R, et al (2018) Shifting the limits in wheat research and breeding using a fully annotated reference genome. Science 361:eaar7191
5. Cobo N, Pflüger L, Chen X et al (2018) Mapping QTL for resistance to new virulent races of wheat stripe rust from two Argentinean wheat cultivars. Crop Sci 14:1–14
6. Dhakal S, Tan C-T, Anderson V et al (2018) Mapping and KASP marker development for wheat curl mite resistance in "TAM 112" wheat using linkage and association analysis. Mol Breed 38:119
7. Qureshi N, Bariana H, Kumran VV et al (2018) A new leaf rust resistance gene Lr79 mapped in chromosome 3BL from the durum wheat landrace Aus26582. Theor Appl Genet 131:1091–1098
8. Yuan C, Wu J, Yan B et al (2018) Remapping of the stripe rust resistance gene Yr10 in common wheat. Theor Appl Genet 131:1253–1262
9. Mourad AMI, El-Wafaa AA, Wegulo S et al (2018) Genome-wide association study for identification and validation of novel SNP markers for Sr6 stem rust resistance gene in bread wheat. Front Plant Sci 9:1–12
10. Marchal C, Zhang J, Zhang P et al (2018) BED-domain-containing immune receptors confer diverse resistance spectra to yellow rust. Nat Plants 4:662–668
11. Mo Y, Vanzetti LS, Hale I et al (2018) Identification and characterization of Rht25, a locus on chromosome arm 6AS affecting wheat plant height, heading time, and spike development. Theor Appl Genet 131:2021–2035
12. Zhai H, Feng Z, Du X et al (2018) A novel allele of TaGW2-A1 is located in a finely mapped QTL that increases grain weight but decreases grain number in wheat (Triticum aestivum L.). Theor Appl Genet 131:539–553
13. Würschum T, Langer SM, Longin CFH et al (2017) A modern green revolution gene for reduced height in wheat. Plant J 92:892–903
14. Demichelis M, Vanzetti LS, Crescente JM et al (2019) Significant effects in bread-making quality associated with the gene cluster Glu-D3/Gli-D1 from the bread wheat cultivar Prointa Guazú. Cereal Res Commun 47:111–122

15. Alomari DZ, Eggert K, von Wirén N et al (2018) Identifying candidate genes for enhancing grain Zn concentration in wheat. Front Plant Sci e9:1313
16. Velu G, Singh RP, Crespo-Herrera L et al (2018) Genetic dissection of grain zinc concentration in spring wheat for mainstreaming biofortification in CIMMYT wheat breeding. Sci Rep 8:13526
17. Thind AK, Wicker T, Šimková H et al (2017) Rapid cloning of genes in hexaploid wheat using cultivar-specific long-range chromosome assembly. Nat Biotechnol 35:793
18. Sánchez-Martín J, Steuernagel B, Ghosh S et al (2016) Rapid gene isolation in barley and wheat by mutant chromosome sequencing. Genome Biol 17:221
19. Helguera M, Rivarola M, Clavijo B et al (2015) New insights into the wheat chromosome 4D structure and virtual gene order, revealed by survey pyrosequencing. Plant Sci 233:200–212
20. Vitulo N, Albiero A, Forcato C et al (2011) First survey of the wheat chromosome 5A composition through a next generation sequencing approach. PLoS One 6:e26421
21. Hernandez P, Martis M, Dorado G et al (2012) Next-generation sequencing and syntenic integration of flow-sorted arms of wheat chromosome 4A exposes the chromosome structure and gene content. Plant J 69:377–386
22. Tanaka T, Kobayashi F, Joshi GP et al (2014) Next-generation survey sequencing and the molecular Organization of Wheat Chromosome 6B. DNA Res 21:103–114
23. Berkman PJ, Skarshewski A, Lorenc MT et al (2011) Sequencing and assembly of low copy and genic regions of isolated Triticum aestivum chromosome arm 7DS. Plant Biotechnol J 9:768–775
24. Berkman PJ, Skarshewski A, Manoli S et al (2012) Sequencing wheat chromosome arm 7BS delimits the 7BS/4AL translocation and reveals homoeologous gene conservation. Theor Appl Genet 124:423–432
25. Šimková H, Svensson JT, Condamine P et al (2008) Coupling amplified DNA from flow-sorted chromosomes to high-density SNP mapping in barley. BMC Genomics 9:294
26. Dean FB, Hosono S, Fang L et al (2002) Comprehensive human genome amplification using multiple displacement amplification. Proc Natl Acad Sci U S A 99:5261–5266
27. Barker DL, Hansen MST, Faruqi AF et al (2004) Two methods of whole-genome amplification enable accurate genotyping across a 2320-SNP linkage panel. Genome Res 14:901–907
28. Matsunaga S, Kawano S, Michimoto T et al (1999) Semi-automatic laser beam microdissection of the Y chromosome and analysis of Y chromosome DNA in a Dioecious plant, Silene latifolia. Plant Cell Physiol 40:60–68
29. Vitharana SN, Wilson GS (2006) Fractionation of chromosome 15 with an affinity-based approach using magnetic beads. Genomics 87:158–164
30. Doležel J, Vrána J, Cápal P et al (2014) Advances in plant chromosome genomics. Biotechnol Adv 32:122–136
31. Heltzen M, Baker S (2019) Illumina. In: AllSeq the sequencing marketplace. http://allseq.com/knowledge-bank/sequencing-platforms/illumina/. Accessed 29 Apr 2019
32. Plöthner M, Frank M, von der Schulenburg J-MG (2017) Cost analysis of whole genome sequencing in German clinical practice. Eur J Health Econ 18:623–633
33. Heltzen M, Baker S (2019) 454 Roche. In: AllSeq the sequencing marketplace. http://allseq.com/knowledge-bank/sequencing-platforms/454-roche/. Accessed 29 Apr 2019
34. Merriman B, R&D Team IT, Rothberg JM (2012) Progress in ion torrent semiconductor chip based sequencing. Electrophoresis 33:3397–3417
35. Heltzen M, Baker S (2019) Ion torrent. In: AllSeq the sequencing marketplace. http://allseq.com/knowledge-bank/sequencing-platforms/ion-torrent/. Accessed 29 Apr 2019
36. Heltzen M, Baker S (2019) SOLiD. In: AllSeq the sequencing marketplace. http://allseq.com/knowledge-bank/sequencing-platforms/solid/. Accessed 29 Apr 2019
37. Rhoads A, Au KF (2015) PacBio sequencing and its applications. Genomics Proteomics Bioinformatics 13:278–289
38. Heltzen M, Baker S (2019) Pacific biosciences. In: AllSeq the sequencing marketplace. http://allseq.com/knowledge-bank/sequencing-platforms/pacific-biosciences/. Accessed 29 Apr 2019
39. https://nanoporetech.com/products/comparison. Accessed 29 Apr 2019
40. Goodwin S, Gurtowski J, Ethe-Sayers S et al (2015) Oxford Nanopore sequencing, hybrid error correction, and de novo assembly of a eukaryotic genome. Genome Res 25:1750–1756
41. Walker BJ, Abeel T, Shea T et al (2014) Pilon: an integrated tool for comprehensive microbial variant detection and Genome assembly improvement. PLoS One 9:e112963
42. Loman NJ, Quick J, Simpson JT (2015) A complete bacterial genome assembled de novo

using only nanopore sequencing data. Nat Methods 12:733–735

43. Schwartz D, Li X, Hernandez L et al (1993) Ordered restriction maps of Saccharomyces cerevisiae chromosomes constructed by optical mapping. Science 262:110–114
44. Tang H, Lyons E, Town CD (2015) Optical mapping in plant comparative genomics. GigaScience 4:3
45. Deschamps S, Zhang Y, Llaca V et al (2018) A chromosome-scale assembly of the sorghum genome using nanopore sequencing and optical mapping. Nat Commun 9:4844
46. Jiao Y, Peluso P, Shi J et al (2017) Improved maize reference genome with single-molecule technologies. Nature 546:524–527
47. Mascher M, Gundlach H, Himmelbach A et al (2017) A chromosome conformation capture ordered sequence of the barley genome. Nature 544:427–433
48. Mayer KFX, Rogers J, Dolezel J et al (2014) A chromosome-based draft sequence of the hexaploid bread wheat (Triticum aestivum) genome. Science 345:1251788
49. Simpson JT, Wong K, Jackman SD et al (2009) ABySS: a parallel assembler for short read sequence data. Genome Res 19:1117–1123
50. Schatz MC, Delcher AL, Salzberg SL (2010) Assembly of large genomes using second-generation sequencing. Genome Res 20:1165–1173
51. Zerbino DR, Birney E (2008) Velvet: algorithms for de novo short read assembly using de Bruijn graphs. Genome Res 18:821–829
52. Li R, Zhu H, Ruan J et al (2010) De novo assembly of human genomes with massively parallel short read sequencing. Genome Res 20:265–272
53. Margulies M, Egholm M, Altman WE et al (2005) Genome sequencing in microfabricated high-density picolitre reactors. Nature 437:376–380
54. Akpinar BA, Biyiklioglu S, Alptekin B et al (2018) Chromosome-based survey sequencing reveals the genome organization of wild wheat progenitor Triticum dicoccoides. Plant Biotechnol J 16:2077–2087
55. Koren S, Walenz BP, Berlin K et al (2017) Canu: scalable and accurate long-read assembly via adaptive k -mer weighting and repeat separation. Genome Res 27:722–736
56. Lu H, Giordano F, Ning Z (2016) Oxford Nanopore MinION sequencing and Genome assembly. Genomics Proteomics Bioinformatics 14:265–279
57. Ribeiro FJ, Przybylski D, Yin S et al (2012) Finished bacterial genomes from shotgun sequence data. Genome Res 22:2270–2277
58. Ye C, Hill CM, Wu S et al (2016) DBG2OLC: efficient assembly of large genomes using long erroneous reads of the third generation sequencing technologies. Sci Rep 6:1–9
59. Yu J (2002) A draft sequence of the rice genome (Oryza sativa L. ssp. indica). Science 296:79–92
60. Du H, Yu Y, Ma Y et al (2017) Sequencing and de novo assembly of a near complete indica rice genome. Nat Commun 8:1–12
61. Li C, Song W, Luo Y et al (2019) The HuangZaoSi maize Genome provides insights into genomic variation and improvement history of maize. Mol Plant 12:402–409
62. Paterson AH, Bowers JE, Bruggmann R et al (2009) The Sorghum bicolor genome and the diversification of grasses. Nature 457:551–556
63. McCormick RF, Truong SK, Sreedasyam A et al (2018) The Sorghum bicolor reference genome: improved assembly, gene annotations, a transcriptome atlas, and signatures of genome organization. Plant J 93:338–354
64. Schnable PS, Page SEEL, Pasternak S et al (2012) The B73 maize genome: complexity, diversity, and dynamics. Science 326:1112–1115
65. International T, Genome B, Consortium S (2012) A physical, genetic and functional sequence assembly of the barley genome. Nature 491:711–716
66. Clavijo BJ, Venturini L, Schudoma C et al (2017) An improved assembly and annotation of the allohexaploid wheat genome identifies complete families of agronomic genes and provides genomic evidence for chromosomal translocations. Genome Res 27:885–896

Chapter 5

Cost-Effective Profiling of *Mutator* Transposon Insertions in Maize by Next-Generation Sequencing

Xinyan Zhang, Meixia Zhao, and Damon Lisch

Abstract

Transposable elements can be highly mutagenic because when they transpose they can insert into genes and disrupt their function, a propensity which has been exploited in many organisms to generate tagged mutant alleles. The *Mutator* (*Mu*) family transposon is a family of DNA-type transposons in maize with a particularly high duplication frequency, which results in large numbers of new mutations in lineages that carry active *Mu* elements. Here we describe a rapid and cost-effective Miseq-based *Mu* transposon profiling pipeline. This method can also be used for identifying flanking sequences of other types of long insertions such as T-DNAs.

Key words Transposon, *Mutator*, Insertion profiling, Next-generation sequencing (NGS), Target site duplication, Terminal inverted repeats

1 Introduction

A large fraction of many plant and animal genomes are made up of transposable elements (TEs) [1], which are pieces of DNA fragments that can "jump" to new positions within a genome. TEs can be divided to two major types: Class I retrotransposons and Class II DNA transposons [2]. Transposition of Class II transposons requires enzymes known as transposases, which are encoded by the autonomous members of a given family of elements. Those elements that do not encode transposases by themselves but are capable of transposing in the presence of transposases are known as nonautonomous elements. Both autonomous and nonautonomous transposable elements produce double strand breaks upon excision and insertional mutations in the host genome following integration [3, 4].

One class of the DNA transposons, the *Mutator* (*Mu*) family of transposons, was first identified in maize and was named for its high mutagenic capability [5]. *Mu* elements transpose by a cut-and-paste mechanism. The transposases recognize their conserved

Luis M. Vaschetto (ed.), *Cereal Genomics: Methods and Protocols*, Methods in Molecular Biology, vol. 2072,
https://doi.org/10.1007/978-1-4939-9865-4_5,

binding sites embedded in ~200-bp terminal inverted repeats (TIRs) found at both ends of each *Mu* element [6]. Integration of the *Mu* elements into the genome creates a characteristic 9-bp target site duplication (TSD) of flanking DNA [7]. High-copy *Mutator* lines have been used for targeted mutagenesis and forward genetic screening by many groups (reviewed by Candela and Hake) [8]. However, cloning the causal mutations for the mutants isolated from randomly mutagenized populations can be technically challenging due to the high copy number of *Mu* transposons. *Mutator* lines have been used in constructing annotated ready-to-order mutant seed libraries to advance maize reverse genetic research. Deep-sequencing-based pipelines have been developed for *Mu* insertion profiling, which involve sequencing amplicon PCR products containing the TIR-flanking junction sequences [9–11]. These pipelines are generally large scale and need to be modified for routine applications at a low cost in laboratories that do not specialize in these methodologies. Here we describe a rapid and cost-effective Miseq-based flanking sequence identification method that is tailored for *Mu* transposon profiling but that is potentially equally useful for other TEs in other species, as well as other types of long insertions such as T-DNAs.

2 Materials

Prepare all reagents at room temperature, unless indicated otherwise. Prepare all solutions using molecular grade water.

2.1 Wideseq Sample Preparation

1. Lysis buffer (for 50 mL) (*see* **Note 1**): 5 mL 1 M Tris–HCl (pH 8, final concentration: 100 mM), 5 mL 0.5 M EDTA (final concentration: 50 mM) (*see* **Note 2**), 5 mL 5 M NaCl (final concentration: 500 mM), 35 mL water.
2. 10% sodium dodecyl sulfate (SDS) solution in water.
3. 5 M potassium acetate (KOAc).
4. Propanol.
5. 70% ethyl alcohol.
6. DNase and 2–10 mg/mL protease-free RNase A solution: Store at −20 °C.
7. Qubit™ dsDNA HS Assay Kit (ThermoFisher).
8. DNA shearing reagents: dsDNA fragmentase (NEB).
9. Gel DNA Recovery Kit (ZYMO RESEARCH).
10. NEBNext® Ultra™ End Repair/dA-Tailing Module (NEB). Store at −20 °C.
11. 3 M sodium acetate (NaAc) (pH 5.2).

12. 20 mg/mL glycogen: Store at 4 °C.
13. Adaptors: DNA oligos synthesized by Integrated DNA Technologies, Inc. (IDT) (*see* **Note 3**).

    adapter1L 5′-GTAATACGACTCACTATAGGGCACGCGT GGTCGACGGCCCGGGCTGCT-3′.

    adapter1S /5PHOS/GCAGCCCGGG/3AmMO/.

    The adaptors were modified from a next-generation library construction protocol developed for identifying T-DNA insertions [12]. Store at −80 °C.
14. 10 mL Oligo annealing buffer: 1× TE (10 mM Tris, pH 8.0, 1 mM EDTA) 9.9 mL, 5 M NaCl 0.1 mL.
15. Blunt/TA Ligase Master Mix (NEB) Store at −20 °C.
16. Silica Bead DNA Gel Extraction Kit (Thermal).
17. Primers for nested PCR: DNA oligos synthesized by Integrated DNA Technologies, Inc. (IDT).

    AP1-out 5′-GTAATACGACTCACTATAGGGCACGCGTGGTC-3′.

    AP1-nest 5′-TGGTCGACGGCCCGGGCTGC-3′.

    Museq1st 5′-TCTGCGTCTCCARAWCAKAGAAGCCAACG-3′.

    Museq2nd 5′-CGCCATGGCCTCCATTTCGTCGAATCCC-3′
18. Kapa Hifi DNA polymerase (Kapa Biosystems). Store at −20 °C.

2.2 Computational Resources and Software

1. High performance computing cluster or personal computer. The software mentioned in the following text can be run using a Unix-like workstation. For Windows users, a terminal emulator can be used.
2. FastQC (http://www.bioinformatics.babraham.ac.uk/projects/fastqc/).
3. Cutadapt (https://cutadapt.readthedocs.io/en/stable/).
4. Bowtie2 (http://bowtie-bio.sourceforge.net/bowtie2/index.shtml) [13].
5. Samtools (http://samtools.sourceforge.net/) [14].
6. IGV (http://software.broadinstitute.org/software/igv/) [15].
7. BEDtools (http://bedtools.readthedocs.io/en/latest/) [16].

3 Methods

This protocol involves three consecutive sections: amplicon-based enrichment of *Mutator* flanking DNAs, wideseq, and data analysis (Fig. 1). The wideseq sample preparation and wideseq data analysis are the two major focuses of our pipeline given that the wideseq service is commercially available and inexpensive.

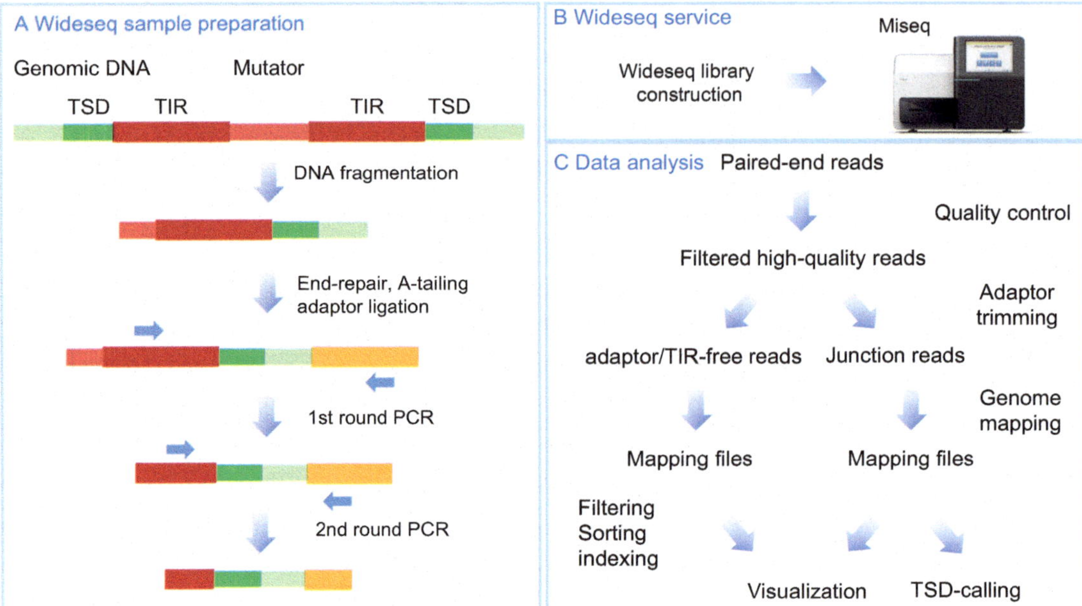

Fig. 1 A schematic view of the wideseq-based *Mutator* profiling pipeline

3.1 Amplicon-Based Enrichment of Mutator Flanking DNA

3.1.1 Genomic DNA Extraction

1. Grind 0.25 g plant tissue in liquid nitrogen to a fine powder in a small mortar.
2. Add 1000 μL of Extraction Buffer and grind again briefly.
3. Pour into a 1.5 mL microfuge tube.
4. Add 120 μL of 10% SDS.
5. Mix well by inverting several times.
6. Keep samples at room temperature for 10 min.
7. Add 300 μL 5 M NaAc. Mix well by inverting several times.
8. Spin for 15 min at top speed in a microfuge.
9. Transfer 600 μL of the supernatant to a new microcentrifuge tube.
10. Add 600 μL of isopropanol.
11. Mix the contents thoroughly by inverting.
12. Spin for 10 min at top speed.
13. Pour off the supernatant and wash the pellet with 500 μL of 70% ethanol.
14. Gently pour off the 70% ethanol and remove the remaining ethanol using a pipettor.
15. Allow DNA to air-dry for 20–30 min.
16. Resuspend the DNA in 50 μL of water.
17. Add 1 μL RNase A to each sample.

18. Incubate at 37 °C for 30 min.
19. Measure the concentration of DNAs using the Qubit dsDNA HS assay. In our hands, the final concentration of DNA using this extraction protocol is 100–200 ng/μL.
20. Optional DNA quality check: Run 3 μL of the DNA sample on a 0.8% agarose gel.

3.1.2 Genomic DNA Digestion

1. Prepare the dsDNA shearing mix (20 μL):

   10 μL DNA (1–2 μg).

   6 μL water.

   2 μL 10× buffer for fragmentase.

   2 μL fragmentase
2. Incubate the mix at 37 °C 30 min.
3. Stop the reaction by adding 5 μL 0.5 M EDTA.
4. Run a 1.5–2% agarose gel.
5. Do a size selection and collect the 500–1500 bp smear of DNA followed by column purification using a Gel DNA Recovery Kit.
6. Elute the DNA with 20 μL elution buffer.

3.1.3 End Repair and A-tailing

1. Prepare the end-repair and A-tailing reaction mix (32.5 μL) as follows:

   17.5 μL purified DNA solution following fragmentation and column purification.

   10.25 μL water.

   3.25 μL buffer (10 X) for end-repair and A-tailing.

   μL end repair enzyme
2. Incubate the mix in a thermal cycler at 20 °C for 30 min, then at 65 °C for 30 min.
3. Add 3.5 μL 3 M NaAc (pH 5.2) to each reaction.
4. Add 1 μL 20 mg/mL glycogen to each sample. Mix well.
5. Add 200 μL ethanol to each sample. Mix well again. Place the mix in the freezer (−20 °C) overnight.
6. Centrifuge at 13000 rpm (16,200 × *g*) for 15 min.
7. Wash the pellets with 75% ethanol. Dry the pellets. Dissolve the DNA in 8 μL of water.

3.1.4 Ligation to Adaptor

1. Prepare the following oligo annealing reaction mix (100 μL):

   25 μL 100 μM adaptor 1 L.

   25 μL 100 μM adaptor 1S.

   50 μL oligo annealing buffer

2. Incubate at 95 °C 1 min, then slowly lower the temperature to 14 °C by −0.1 °C/s.
3. Aliquot 5 μL (adaptor concentration 25 μM) to 20 PCR tubes.
4. Store the adaptors at −80 °C.
5. Perform Qubit assays to measure the concentration of purified DNA insert following end-repair/A-tailing.
6. Based on the concentrations of insert DNA and adaptor, calculate the volumes of adaptors (X μL) and insert DNA (Y μL) for ligation. A 300:1 molar ratio for adaptor: insert is recommended (*see* **Note 4**).
7. Prepare the ligation mix (15 μL):

   X μL adaptor.

   Y μL end-repaired and A-tailed insert DNA.

   μL TA ligation master mix (2×)
8. Incubate the mix at 16 °C for 3 h, perform bead-purification for ligation products (Silica Bead DNA Gel Extraction Kit) following the manufacturer's instructions.
9. Elute DNA with 60 μL of water.

3.1.5 Nested PCR

1. Prepare the first nested PCR reaction mix (15 μL):

   6 μL ligation product.

   4.2 μL water.

   3 μL buffer (5×) for Kapa Hifi.

   0.5 μL primer Museq1st (10 μM).

   0.5 μL primer AP1-out (10 μM).

   0.5 μL dNTPs (10 mM each).

   0.3 μL Kapa Hifi DNA polymerase
2. Mix gently and centrifuge briefly to collect droplets.
3. Run the following PCR program to amplify the first nested PCR reaction mix: 95 °C 3 min; 14 cycles (98 °C 20 s, 60 °C 15 s, 72 °C 45 s); 72 °C 3 min.
4. Prepare the second nested PCR reaction mix (120 μL):

   2 μL first-round PCR product.

   80 μL water.

   24 μL buffer (5×) for Kapa Hifi.

   4 μL primer Museq2nd (10 μM).

   4 μL primer AP1-nest (10 μM).

   4 μL dNTPs (10 mM each).

   2 μL Kapa Hifi

5. Mix gently and centrifuge briefly to collect droplets.
6. Run the following PCR program to amplify the second nested PCR reaction mix: 95 °C 3 min; 14 cycles (98 °C 20 s, 60 °C 15 s, 72 °C 45 s); 72 °C 3 min.
7. Run all the PCR products (120 μL) on a 1.5–2% agarose gel. Select the 500–1500 bp smear region followed by column purification. 8-Elute DNA using 20 μL of the elution buffer.

3.2 Wideseq-Based Sequencing of the Purified DNA Sample

The wideseq service for sequencing the purified DNAs in the above is provided at the Purdue Genomics Core Facility. Wideseq is an inexpensive Next Generation Sequencing (NGS) service which involves the construction of NGS libraries using a transposase-based methodology (Nextera) and miseq-based sequencing. For details see the website of Purdue Genomics Core Facility (https://www.purdue.edu/hla/sites/genomics/wideseq-2/). In a typical wideseq run for one *Mutator*-profiling sample, 20,000 (sometimes up to 50,000) pair-ended reads (2 × 150 bp) are produced (*see* **Note 5**) which are enough for confident identification of the flanking sequences of *Mu* elements in a maize genome with a *Mutator* copy number as high as 100 (*see* **Note 6**) (Table 1).

3.3 Mapping Wideseq Reads and Identifying of Mutator TSDs

3.3.1 Quality Control

We recommend checking the quality of the wideseq sequencing data using FastQC before moving on to the next steps. This step is optional if the wideseq service provides the quality control reports together with the filtered reads based on read quality.

3.3.2 Trim Adaptors and TIR Sequences

1. Trim adaptors by running the cutadapt program using parameter "-g TGGTCGACGGCCCGGGCTGCT --overlap 7 -G TGGTCGACGGCCCGGGCTGCT --overlap 7". This step outputs sequence files with the adaptor trimmed: sample_cutadapt_R1.fastq, sample_cutadapt_R2.fastq.
2. To simplify the TSD calling following genome mapping, pair-ended sequence files (read 1 and read 2) with TSD sequences in read 1 are created during TIR-trimming (*see* **Note 7**):

Table 1
Summary of wideseq-based *Mutator* profiling in samples of two sibling individuals from a *Mutator* stock

	Sample 1	Sample 2
Paired-end reads	145,377	89,461
Mapped concordant adaptor/TIR-free reads	98,220	60,049
Mapped junction reads	3515	3041
Background *Mu* elements in B73	28	28
New insertions	33	47

First, run the cutadapt program with sample_cutadapt_R1.fastq sample_cutadapt_R2.fastq using parameter "-g ATAATGGCAATTATCTC --overlap 7 --discard-untrimmed". Output two files: R1_juncF.fastq and R2_juncF.fastq;

Second, repeat adaptor trimming using sample_cutadapt_R2.fastq sample_cutadapt_R1.fastq as input files. Output files: R1_juncR.fastq and R2_juncR.fastq.

Third, combine reads in files R1_juncF.fastq and R1_juncR.fastq into a new file sample_junction_R1.fastq. Combine reads in files R2_juncF.fastq and R2_juncR.fastq into a new file: sample_junction_R2.fastq. The paired files "sample_junction_R1.fastq; sample_junction_R2.fastq" are to be mapped on the genome for identifying TSD coordinates.

3. Prepare reads for genome mapping in order to visualize adaptor/TIR-free reads by consecutively running cutadapt with parameters "-g TGGTCGACGGCCCGGGCTGCT --overlap 7 -G TGGTCGACGGCCCGGGCTGCT --overlap 7" and "-g ATAATGGCAATTATCTC -G ATAATGGCAATTATCTC --overlap 7", respectively. Output all_clean reads (free of adaptor and TIR sequences) in files: sample1_r1_allclean.fastq and sample1_r2_allclean.fastq.

3.3.3 Map Reads to the Genome

The maize genome B73 RefGen_v4 (Zm-B73-REFERENCE-GRAMENE-4.0) can be downloaded from ftp://ftp.ensemblgenomes.org/pub/plants/release-39/fasta/zea_mays/dna/.

1. Run the bowtie2-build program to build a Bowtie2 large index from the genome sequence file (parameter: --large-index).
2. Run bowtie2 to separately map the junction reads and all trimmed reads to the genome using parameter "--very-sensitive –local" (*see* **Note 8**).
3. Do quality filtering and sam-to-bam conversion with the alignment output (".sam" file) using samtools (parameters: "samtools view -Sh -q 2", "samtools view -S -b").
4. Run samtools again to sort and index the bam files in order to visualize all mapped adaptor/TIR-free reads and junction reads (parameters: "samtools sort -l 5"; "samtools index -b").

Indexed ".bam" file can be visualized using the software IGV.

3.3.4 Extraction of the TSD Coordinates

Because the *Mutator*-specific primers read outward from both TIRs, TIR-trimmed junction reads containing the TSD sequences flanking both TIRs of a given *Mu* element insertion can be aligned, with 9-bp of overlapping sequence that are the TSDs (Fig. 2).

1. To extract the TSD coordinates, convert the filtered mapping file to a ".bed" file by running the sam2bed program (BEDtools).

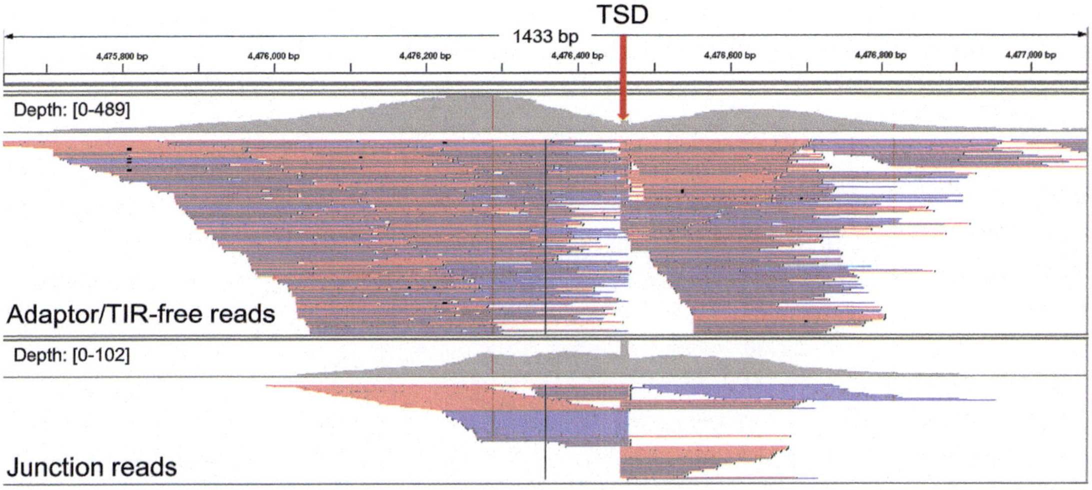

Fig. 2 Visualization of adaptor/TIR-free reads (upper) and junction reads (bottom) mapped to a representative *Mutator* insertion site using the IGV software. The TSD sequence (red arrow) shows a higher mapping depth when visualizing junction reads, as all of these reads contain the TSD sequences at the ends

2. Separately extract two subsets of aligned reads mapped in the forward and reverse directions and write them into two files.
3. Take both the start and the end coordinates for each TIR-trimmed junction read and then use the following rules to obtain a TSD coordinate.

   For each read aligned in forward direction with a START coordinate "m" in the ".bed" file, "m + 1" and "m + 9" are the start and end coordinates of one TSD hit.

   For each read aligned in reverse direction with an END coordinate "n" in the ".bed" file, "n−8" and "n" are the start and end coordinates of one TSD hit (*see* **Note 9**).
4. Create a unique TSD coordinate list by combining all TSD coordinates obtained following the above rules and then remove duplicates. The junction read counts matched to each TSD in forward and reverse directions are summarized separately which are used for further filtering for confident new *Mu* elements as well as *Mu* elements in the B73 background described below.
5. To find confident TSD coordinates, first screen the TSD coordinate list and reserve those with at least one junction read aligned to both flanks of a given *Mu* insertion. At a subset of TSD sites, abundant junction reads (>5 reads) are mapped to only one flank of a given insertion. This is usually the result of two preexisting TIR-TSD junctions associated with background *Mu* elements in the B73 genome, which are separated by each other by the endogenous B73 *Mu* element (Table 2). In contrast, new insertions will generate new TSDs that are not separated by any B73 sequence. The background *Mutator* elements in B73 can often be identified by screening 40 kb

Table 2
Representative *Mutator* insertions identified by this pipeline

			Sample 1		Sample 2		
Chr	TSD_start	TSD_end	count_rev	count_fwd	count_rev	count_fwd	Note
1	269,969,410	269,969,418	49	0	47	0	Background paired
1	269,974,224	269,974,232	0	57	0	47	Background paired
2	26,733,220	26,733,228	21	0	10	0	Background paired
2	26,735,143	26,735,151	0	16	0	13	Background paired
2	142,525,770	142,525,778	64	0	47	0	Background paired
2	142,527,884	142,527,892	0	37	0	30	Background paired
4	30,459,834	30,459,842	29	24	27	21	New insertion
6	65,976,107	65,976,115	26	14	0	0	New insertion
7	138,208,558	138,208,566	0	0	25	23	New insertion
8	15,546,546	15,546,554	5	10	0	0	New insertion
8	97,202,190	97,202,198	0	0	4	18	New insertion
8	124,608,977	124,608,985	0	0	35	17	New insertion
8	162,944,671	162,944,679	38	17	0	0	New insertion
8	165,872,916	165,872,924	0	0	9	17	New insertion
8	170,724,891	170,724,899	28	30	0	0	New insertion
9	113,853,775	113,853,783	35	34	19	17	New insertion

A list of confident new insertions and B73-background *Mutator* insertions with 9-bp TSDs are shown. The background insertions are aligned with paired split peaks

windows neighboring such TSD sites for paired TIR-TSD junction sites (*see* **Note 10**).

The protocol described here provides a rapid and cost-effective way to profile transposons and known insertion sequences with single-nucleotide resolution (*see* **Note 11**). By replacing the *Mutator*-specific primers with other insertion-sequence-specific primers, this pipeline can be potentially adapted for identification of the flanking sequences of known genomic insertions such as T-DNAs and other transposon families.

4 Notes

1. This prep can be scaled up or down.
2. Prepare 0.5 M solution (pH 8) in double distilled water. Use the same solution for terminating DNA fragmentation reaction in Sect. 3.1.2.

3. One of the two oligos is modified. Blocking the 3′ terminus improves the specificity of the PCR. The adaptor sequences and PCR primers can be replaced to meet special needs such as T-DNA profiling. PCR adapters/primers should NOT contain any Illumina library adapter sequences in them, because only DNAs not containing Illumina library adapter sequences are compatible with the post-Nextera PCR reaction in the wideseq assay.
4. Molar ratios between 100:1 and 500:1 (adaptor: insert) have worked in our hands.
5. Submit the DNA sample for the wideseq service as a "library" type.
6. DNA is fragmented during the wideseq library construction and thus, the number of TIR-containing amplicons is underestimated in the sequencing results.
7. Given that the *Mu*-specific primers anneal to both ends of each *Mu* elements and prime outward reading from the ends of TIRs, each *Mu* element insertion is defined by two amplicons: an upstream amplicon derived from the left *Mu* TIR-flanking junction and a downstream amplicon from the right *Mu* TIR-flanking junction. On the two flanks of each *Mu* element, the TIR-flanking junction sequences map to opposite strands. TIR-flanking junction reads from the upstream amplicons are aligned to negative strands of the genome, while those from the downstream amplicons are aligned to positive strands. Since amplicons are sequenced in both forward and reverse directions in wideseq, the directions of each amplicons need to be unified to detect TSD. To do so, the names of read1 and read 2 are simply exchanged during TIR-trimming for those amplicons sequenced in a reverse direction from adaptor to TIR (the TSD seen in read 2). The initial 9 nucleotides in read 1 are where the TSD sequences are locate in each adjusted read pair. This adjustment step is unique to wideseq analysis and is necessary for TSD calling algorithm used in this pipeline, but it is optional for identifying flanking sequences of other types of insertions like T-DNAs.
8. The TSDs but not the TIR sequences remain because of TIR-trimming of these mapped junction reads.
9. In this step a 0-based coordinate from ".bed" file is converted to a 1-based TSD coordinate.
10. In our typical "high-copy" *Mutator* line, 50–100 confident TSDs can be identified, including 20–30 background *Mutator* elements in the B73 genome (Table 1). The paired TSDs associated with background elements are usually separated by an interval of 1–35 kb, so a 40 kb window can be used to filter for these elements. Using background *Mu* elements as internal

controls, one can estimate the efficacy of the *Mutator* profiling pipeline in identifying novel *Mu* elements. In our hands, we can detect all of the background *Mu* elements in B73 that would be expected to amplify using our *Mu*-specific primers with high reproducibility (Table 1).

11. In our hands, one can process up to 24 samples at a time for 40–50 US dollars for each sample including the wideseq service fee. It takes 24–48 hours to prepare the DNAs for wideseq while a routine wideseq service can be completed in 2–4 weeks.

Acknowledgments

This work was supported by the National Science Foundation Grant DBI-1237931 and Purdue Startup Funds to D.L.

References

1. Feschotte C, Pritham EJ (2007) DNA transposons and the evolution of eukaryotic genomes. Annu Rev Genet 41:331–368
2. Lisch D (2013) How important are transposons for plant evolution? Nat Rev Genet 14:49–61
3. Hedges DJ, Deininger PL (2007) Inviting instability: transposable elements, double-strand breaks, and the maintenance of genome integrity. Mutat Res 616:46–59
4. Liu K, Wessler SR (2017) Transposition of Mutator-like transposable elements (MULEs) resembles hAT and Transib elements and V(D)J recombination. Nucleic Acids Res 45:6644–6655
5. Robertson DS (1978) Characterization of a mutator system in maize. Mutation Res 51:21–28
6. Benito MI, Walbot V (1997) Characterization of the maize mutator transposable element MURA transposase as a DNA-binding protein. Mol Cell Biol 17:5165–5175
7. Lisch D (2002) Mutator transposons. Trends Plant Sci 7:498–504
8. Candela H, Hake S (2008) The art and design of genetic screens: maize. Nat Rev Genet 9:192–203
9. McCarty DR, Settles AM, Suzuki M et al (2005) Steady-state transposon mutagenesis in inbred maize. Plant J 44:52–61
10. Williams-Carrier R, Stiffler N, Belcher S et al (2010) Use of Illumina sequencing to identify transposon insertions underlying mutant phenotypes in high-copy mutator lines of maize. Plant J 63:167–177
11. McCarty DR, Latshaw S, Wu S et al (2013) Mu-seq: sequence-based mapping and identification of transposon induced mutations. PLoS One 8(10):e77172
12. O'Malley RC, Alonso JM, Kim CJ et al (2007) An adapter ligation-mediated PCR method for high-throughput mapping of T-DNA inserts in the Arabidopsis genome. Nat Protoc 2:2910–2917
13. Langmead B, Salzberg SL (2012) Fast gapped-read alignment with Bowtie 2. Nat Methods 9:357–359
14. Li H, Handsaker B, Wysoker A et al (2009) The sequence alignment/map format and SAMtools. Bioinformatics 25:2078–2079
15. Robinson JT, Thorvaldsdottir H, Winckler W et al (2011) Integrative genomics viewer. Nat Biotechnol 29:24–26
16. Quinlan AR, Hall IM (2010) BEDTools: a flexible suite of utilities for comparing genomic features. Bioinformatics 26:841–842

Chapter 6

Relative Expression Analysis of Target Genes by Using Reverse Transcription-Quantitative PCR

Rocío Liliana Gómez and Lorena Noelia Sendín

Abstract

Real-time PCR is a powerful technique used for quantification of defined nucleic acid sequences. Numerous applications of this method have been described including: gene expression studies, diagnosis of pathogens, and detection of genetically modified organisms or mutations. Here, we describe a simple and efficient protocol to determine gene expression in cereals, based on real-time PCR using SYBR® Green dye. This technique provide an inexpensive alternative, since no probes are required, allowing for the quantification of a high number of genes with reduced cost.

Key words Real time, Gene expression, Plants

1 Introduction

Real-time or quantitative reverse-transcription polymerase chain reaction (RT-qPCR) is a sensitive, rapid, and specific technique commonly used to detect and measure small quantities of nucleic acid [1]. RT-qPCR is used for different purposes, for example in cereals this technique was used for expression quantitation of genes either own plant or foreign [2–4]; another use is detection of contaminating microorganisms [5]; and assay validation experiments of microarrays or RNAseq among others [6, 7].

RT-qPCR allows a direct measurement of amplification during the log linear phase of the PCR reaction via the incorporation of a fluorescent probe in the PCR reaction mix and the use of a thermocycler equipped with an optical sensor for fluorescence [8]. The principle of the method is that as the specific-sequence PCR product is accumulated, the fluorescence of the reaction increases until it exceeds a threshold set by the operator in a determinate cycle number. This is called the threshold cycle (Ct), which is inversely proportional to the initial number of cDNA target molecules. Thus, the smaller the Ct values (i.e., fewer number of cycles to

Luis M. Vaschetto (ed.), *Cereal Genomics: Methods and Protocols*, Methods in Molecular Biology, vol. 2072, https://doi.org/10.1007/978-1-4939-9865-4_6,

reach detectable product), the greater the number of initial target molecules.

By comparing the Ct value of the target gene to that of an internal control gene (a constitutively expressed gene, e.g., actin), it is possible to determine its relative abundance in a starting population of mRNA [9, 10]. This method utilizes intercalating agents or fluorogenic probes to sense amplified products. Depending on the fluorescent used, the methods are classified in specific or nonspecific [11]. The first one is based on probes harboring both a fluorescent dye and a quencher molecule, which hybridize with a specific region of the target sequence [12]. The process is based on the transfer of fluorescent energy by resonance (FRET) between the two molecules. This principle consists in that a high energy molecule near a low energy molecule (quencher), promotes an energy transfer and there will be no fluorescence emission. Once these molecules are separated, the fluorescence is emitted, which is captured by the reader of the device [13]. There are several detection systems based on this method used in plant studies, such as Taqman™, Molecular Beacons, and Scorpions™. The TaqMan™ system consists in the detection of the PCR product by using an oligonucleotide probe labeled with a fluorophore and a fluorescence quencher that, upon binding to the DNA template, undergoes quencher removal by the 5′–3′ exonuclease activity of Taq DNA polymerase [14]. In this sense, the amount of amplicon produced is directly proportional to the amount of probe that is released [10]. Molecular beacon is another option, which contains a target recognition loop flanked by a hairpin with a fluorophore and a quencher on the opposing ends. Specific binding to a PCR product opens the hairpin, separating the fluorophore and the quencher, thus producing fluorescence [15]. Scorpion is a beacon variant that is a PCR primer with a hairpin structure at the 5′ end containing a fluorophore and a quencher. When the probe binds to a specific region of the target DNA, the oligonucleotide attached to the quencher is released, folding the fluorochrome-labeled probe over the target releasing the fluorescence [16]. Under fast thermal cycling conditions, Scorpion primers may outperform TaqMan or beacon probes [17].

The nonspecific method uses intercalating molecules. The most common is the SYBR Green I molecule, because of its high affinity for double-stranded DNA and because it is an affordable alternative [18]. Despite hybridization probes offer the advantage of target sequence specificity, a specific probe is required; in that sense, the synthesis of fluorescence-labeled probes becomes very expensive.

Therefore in this chapter we describe the analysis of gene expression by using SYBR Green®, since it offers a great benefit allowing the quantification of a high number of genes, it is economically convenient, there is no need to design probes and the reaction conditions are easy to optimize. However, a weakness of this technique is that SYBR Green® is uniformly integrated into every amplicon, measuring both nonspecific and specific amplified products. Nevertheless, the nonspecific fluorescence can be avoided optimizing the amplification by performing a melting curve at the end of the reaction [19].

2 Materials

2.1 Tissue Collection

The cereal tissues (*see* **Note 1**) are collected from the field in an aluminum bag placed in a container with ice and immediately frozen with liquid nitrogen until use (*see* **Note 2**).

2.2 RNA Extraction

1. Refrigerated microcentrifuge and rotor capable of reaching up to 12,000 g.
2. Sterile polypropylene microcentrifuge tubes (1.5 and 2.0 mL) (*see* **Note 3**).
3. Sterile plastic filter tips (1–20 μL; 10–200 μL; 100–1.000 μL Eppendorf tips) (*see* **Note 4**).
4. P20, P200, and P1000 micropipettes (*see* **Note 4**).
5. Gloves (*see* **Note** 5).
6. Mortars, pestles.
7. RNA isolation reagents. RNA extraction was developed by using TRIzol® (Ambion life technologies—Carlsbad, California, USA) following the manufacturer's instructions for Tissues samples. Extra materials needed: chloroform, isopropyl alcohol, ethanol 75% (in DEPC-treated water), RNase-free water.
8. Liquid nitrogen.
9. 10% bleach solution.

2.3 RNA Monitoring

1. Agarose.
2. Gel running buffer (Loading 6× and GelRed).
3. RNase-free water (DEPC-treated water).
4. UV/VIS spectrophotometer.
5. Quartz cuvettes.
6. Cleaning RNA reagent. DNaseI, RNase-free kit (Thermo Scientific). Preparation of DNA-free RNA was performed by using DNaseI, RNase-free kit (Thermo Scientific) following the manufacturer's instructions for tissues samples.

2.4 cDNA Synthesis

1. cDNA synthesis reagents. RevertAid Reverse Transcriptase (RT) kit (Thermo Scientific). First strand cDNA synthesis for real time qPCR was achieved by using RevertAid Reverse

Transcriptase (RT) kit (Thermo Scientific) following the manufacturer's instructions for RNA samples,

2. Oligo $(dT)_{15}$ solution (Promega).
3. 10 mM dNTP mix.
4. RNase inhibitor (Ribolock—Thermo Fisher scientific).
5. Heating block.

2.5 RT- qPCR Assay

1. Clean lab coats (*see* **Note 6**).
2. Tips and micropipettes for PCR (*see* **Note 4**).
3. Polypropylene tubes for reagent storage and master mix preparation (1.5 mL microcentrifuge tubes).
4. RT-qPCR plate (48 or 96 wells) and flat caps (Bio-Rad, California, USA).
5. Real time equipment.
6. Software for primers sequence design.
7. iQ™ SYBR® Green supermix (Bio-Rad, California, USA),
8. cDNA samples,
9. Microplate Centrifuge

3 Methods

3.1 Material Collection

1. Cut the cereal leaves with a clean pair of scissors; the tissues must be rapidly frozen in liquid nitrogen.
2. Put the leaves in mortars and grind thoroughly with a pestle (*see* **Note 7**).
3. Transfer the ground tissue to a liquid-nitrogen-cooled 1.5 mL microcentrifuge tube (*see* **Note 8**).

3.2 RNA Preparation

3.2.1 RNA Isolation

1. Before starting, clean the working area and equipment with 10% bleach solution for an RNase-free environment. The extraction is one of the most important steps and should be done carefully. It is important to use the correct amount of starting material; do not surpass 100 mg of ground tissue per mL of TRIzol reagent (*see* **Note 9**). Proceed following the manufacturer's instructions.
2. Perform all centrifugation steps at 4 °C in microcentrifuge.
3. After finishing the extraction, the RNA extracted must be refrigerated at −70 °C until use (*see* **Note 10**).

3.2.2 RNA Monitoring

1. In order to determine the RNA quantity and quality, mix 4 μl of RNA sample with 996 μl RNase-free water.

2. Measure the absorbance at 230, 260, and 280 nm, using RNase-free water as zero to the spectrophotometer. An abs_{260} = 1 corresponds to 40 μg/mL. Use the formula A_{260} × dilution × 40 = μg RNA/mL to calculate RNA concentration [20] (*see* **Note 11**).
3. With the purpose of checking the integrity of RNA, a denaturing agarose gel electrophoresis should be performed (*see* **Note 12**).
4. Mix 4 μL RNA sample with 1 μL Loading 6× and 1uL GelRed (1/100) and carry out the running. The ribosomal RNAs (18S and 28S) should appear as well defined bands; if the ribosomal bands are not sharp, it means that the samples have undergone degradation.
5. It is possible that the extracted RNA contains DNA contamination (Fig. 1) (*see* **Note 13**). In order to eliminate DNA we use the enzyme DNase I according to the manufacturer's instructions (*see* **Note 14**).

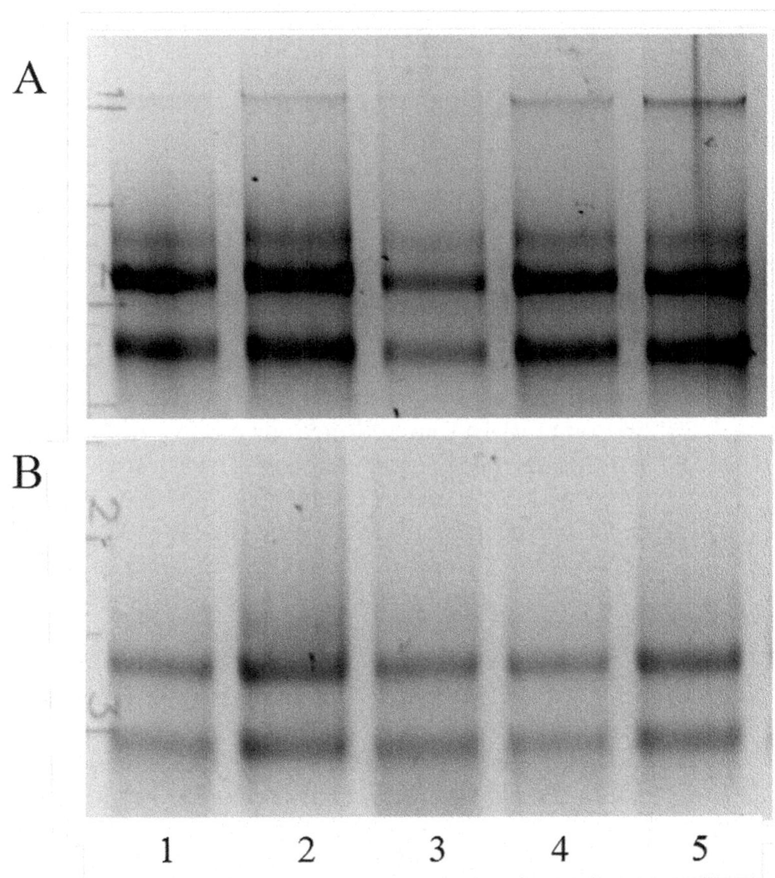

Fig. 1 Monitoring of the integrity and quality of RNA. Denaturing agarose gel electrophoresis of (**a**) RNA samples without DNase treatment and (**b**) RNA samples with DNase treatment

3.3 Real-Time qPCR (RT-qPCR)

In order to detect the expression of specific genes, real time quantitative PCR (RT-qPCR) is performed. RT-qPCR use DNA as template, which is obtained by reverse transcription of RNA

3.3.1 cDNA Synthesis

1. First strand cDNA is synthesized by using RevertAid Reverse Transcriptase (RT) kit following the manufacturer's instructions for RNA samples.
2. After finishing cDNA synthesis, store it at −20 °C until use (*see* **Note 15**).

3.3.2 Primers Design

To facilitate primers design for RT-qPCR, use specific software (*see* **Note 16**). The software shows a list of possible primers, so choose according to the following requirements:

- An amplicon size between 70 and 150 bp.
- An optimal primer length of 20 bases.
- A GC percentage between 30 and 80%.
- Do not use consecutive identical nucleotides. Try to avoid areas with more than 4 G's.
- The last 5 nucleotides of the 3′ end should not have more than 2 G's and/or Cs.
- The Tm should be between 58 and 60 °C.

Resuspend all primers in sterile water at the concentration of 100 μM and storage them at −20 °C.

3.3.3 RT-qPCR Assay Conditions

Gene expression data generated by real-time PCR can be analyzed by two common methods: absolute quantification and relative quantification [21]. Here we describe the $2^{-\Delta\Delta Ct}$ method, which is used for relative quantification [22].

The gene expression can be relative to two parameters. The first one involves an endogenous control or housekeeping (a constantly expressed reference gene). The housekeeping gene is used to normalize the data in order to compensate the amount of DNA and other variants in each sample (*see* **Note 17**). The housekeeping depends on the crop that is analyzed; some of them are GAPDH, F-Box and puroindoline-b in maize, soybeans and wheat respectively [23–25].

The second parameter can be relative to a particular condition (e.g., a nontreated control, a time point zero, or healthy individuals) [26].

To determine the level of expression, the threshold cycle (Ct) values corresponding to each sample are needed. In that sense, the Ct (raw data) values should be measured for:

- The housekeeping gene: control and experimental conditions.
- The gene of interest: control and experimental conditions.

Table 1
Example of raw data values thrown by the Real time thermal cycler software

	Housekeeping gene			Gene of interest		
Sample	**Ct1**	**Ct2**	**Ct3**	**Ct1**	**Ct2**	**Ct3**
Control 1	17.76	17.78	17.75	30.22	30.25	30.20
Control 2	17.98	17.95	17.96	30.05	30.12	30.07
Control 3	17.86	17.88	17.85	30.15	30.16	30.19
Treated 1	18.25	18.24	18.22	26.32	26.35	26.33
Treated 2	18.12	18.10	18.05	27.05	26.95	26.99
Treated 3	18.20	18.23	18.18	26.55	26.61	26.57

For example, in the Table 1, the values of each sample (control and treated) have been represented. We suggest the use of at least three technical replicates per sample (Ct1, Ct2 and Ct3) when performing this technique (*see* **Note 18**). Control 1, Control 2 and Control 3 are biological replicates.

To establish the level of gene expression by using the double delta ct analysis, the following steps should be conducted:

1. Calculate the average of the Ct values for the technical replicates for the housekeeping gene and the gene of interest in the experimental and control conditions.

$$\text{Ct Average}=(\text{Ct1}+\text{Ct2}+\text{Ct3})/\text{Numbers of replicates}$$

$$\text{For example:Ct Average}(\text{“Control 1”})=(17.76+17.78+17.75)/3$$

2. Use the average Ct values obtained above in order to calculate delta Ct (ΔCt) for each sample [26] (*see* **Note 19**).

$$\Delta\text{Ct}=\text{Ct}(\text{gene of interest})-\text{Ct}(\text{housekeeping gene})$$

$$\Delta\text{Ct}(\text{“Control 1”})=30.22\text{-}17.76$$

3. Select the sample or group of samples that will be used as reference (nontreated control, a time point zero or healthy individuals) in order to calculate the delta-delta Ct ($\Delta\Delta$Ct) values for all the samples (*see* **Note 20**).

$$\Delta\Delta\text{Ct}=\Delta\text{Ct}(\text{Treated})-\Delta\text{Ct}(\text{Control})$$

4. In order to know the fold gene expression, use the formula described below:

$$\text{Fold gene expression} = 2^{\wedge-(\Delta\Delta Ct)}$$

5. Previous to begin the assay, design the RT-qPCR plate strategically in order to avoid mistakes (Fig. 2) (*see* **Note 21**).
6. Thaw and homogenize all the reaction components; iQ™ SYBR Green SuperMix (*see* **Note 22**), forward and reverse primers and sterile water. Prepare the master mix by adding iQ™ SYBR Green SuperMix (2×), primers (50–900 nM depending on the optimized concentration (Fig. 3) (*see* **Notes 23** and **24**)) and water following the manufacturer's instructions. Aliquot equal volumes in the wells of the qPCR plate (*see* **Note 25**) and add cDNA (0.001–100 ng) (*see* **Note 26**) in a separated area. Seal wells with flat caps.
7. For a correct reading, be sure to mix all the components of the reaction. Centrifuge the plate to remove the bubbles and collect everything at the bottom of the tube.

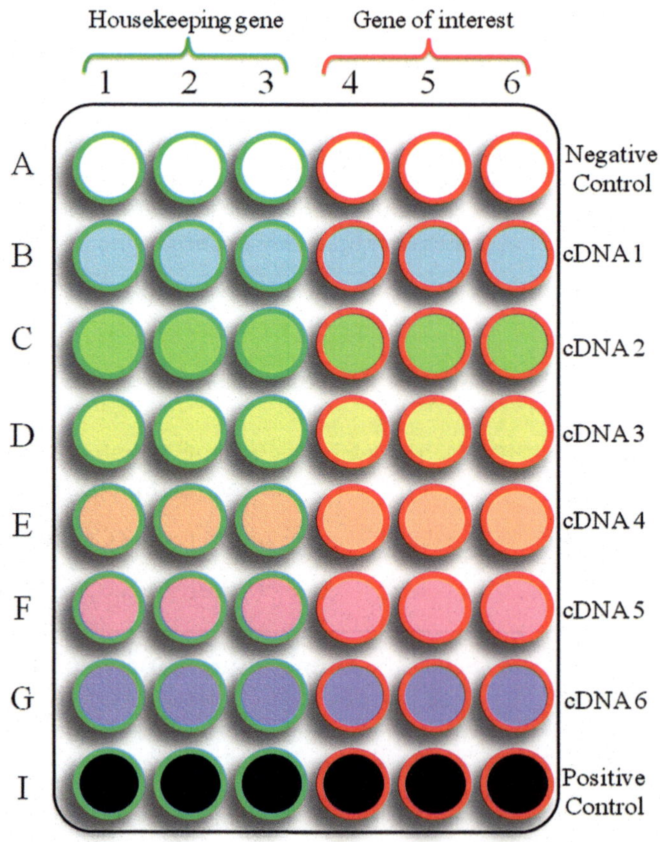

Fig. 2 Illustration of the distribution of PCR plate. Primers are loaded in the columns (border color), whereas cDNAs are loaded across row (fill color)

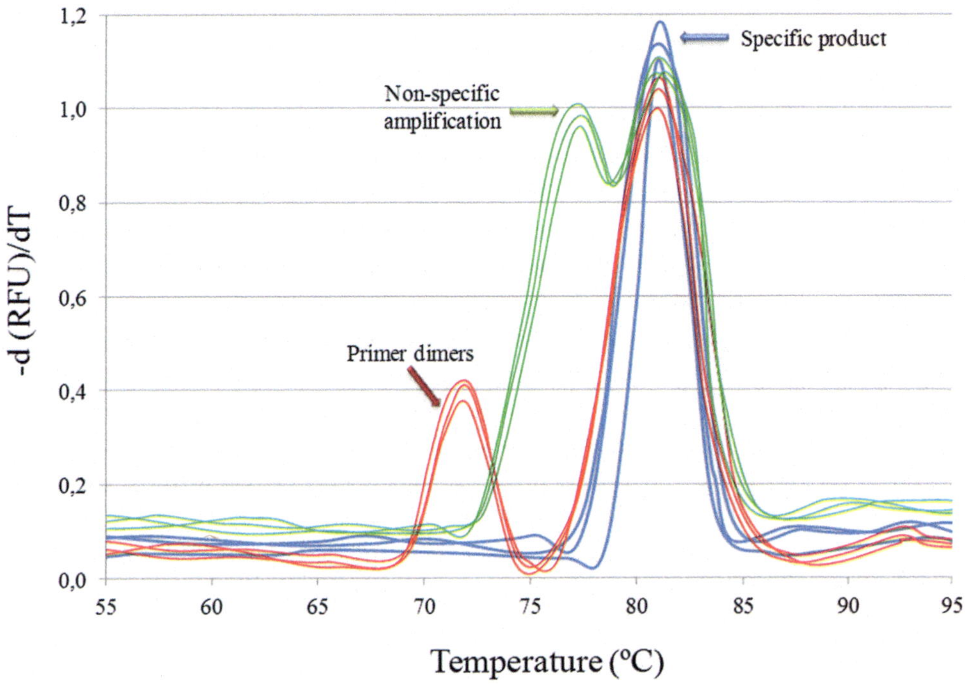

Fig. 3 Diagram of the use of melting curve analysis to discern between specific and nonspecific PCR products

8. The thermal cycling protocol consisted of 1 cycle of polymerase activation and DNA denaturation at 95 °C for 3 min; and 40 cycles of amplification containing the following steps: 10 s of denaturation at 95 °C, 35 s of annealing/extension step at 58–60 °C, depending on the optimum Tm of the primer. The reactions were run and analyzed on a real-time PCR thermal cycler.

4 Notes

1. We use leaves instead of seed endosperm because it contains very high levels of starch, which leads to low yields and poor quality of RNA extraction [27].
2. Tissues can be stored at −70 °C for several months or used in the moment. Add as much liquid Nitrogen to the samples as they need; do not allow tissues to thaw. The process should be carried out as quickly as possible.
3. The microcentrifuge tubes must be sterilized twice.
4. It is important to use separated and confined areas for each step (sample preparation, PCR setup, PCR amplification) in

order to avoid cross-contamination with amplicons that lead to the production of false positives. Do not transfer tips and micropipettes from one area to another.

5. Always wear disposable gloves, and change them whenever you suppose that they are contaminated or in each working area in order to decrease the risk of RNA degradation.
6. Do not use the same lab coat while amplification of PCR products is been performed or when the sample is been prepared.
7. Mortars, pestles and microcentrifuge tubes can be refrigerated before starting. This will help the tissue to not lose its cold temperature.
8. Do not permit tissues to thaw while handling prior to add the TRIzol reagent. This procedure should be carried out as quickly as possible.
9. To avoid thawing while the tissue is weighed, weigh another amount of ground tissue beforehand and see how far the tube is filled.
10. Aliquot RNA in another RNase-free tube in order to check the quality and quantity of RNA extracted. This will prevent the contamination with ribonucleases while handling for check it.
11. The absorbance at 260 should be higher than 0.15 in order to decrease the measurement error. An $abs_{260/280}$ ratio between 1.7 and 2.1 indicates a clean RNA, values out of this range indicates protein contamination. An $abs_{260/230}$ ratio about 2 indicates a clean RNA, values out of this range indicates carbohydrate contamination.
12. Prepare a new gel with RNase-free water; wash the gel box with detergent and pure ethanol.
13. Traces of DNA can be detected on the agarose gel (on top of the gel) (Fig. 1a). DNA must be eliminated since it can interfere in the following gene expression analysis through real-time RT-qPCR.
14. In most cases, the DNase cleaning decreases the amount of total RNA in the sample (Fig. 1b). Be carefully, do not exceed the indicated amount of DNaseI recommended; this could degrade the RNA. After finishing the cleaning, another electrophoresis gel is run in order to check the absence of DNA traces.
15. Before storing the cDNA, aliquot in different sterile microcentrifuge tubes in order to avoid its degradation by thawing and freezing repeatedly.
16. The IDT software™ allows you to design your own primers and probes for Real time qPCR using SYBR Green dye.

17. A good housekeeping should not have large variations of gene expression among the samples analyzed. Its expression must not be related to the treatment to be evaluated.
18. Three technical replicates per sample allow identifying if something odd happened in a well. It is expected that if the experiment was performed correctly, the three replicates should have similar values.
19. The Delta Ct (ΔCt) is the difference in Ct values for the gene of interest and the housekeeping gene for a given sample. This is to normalize the gene of interest to a gene which is not affected by your experiment.
20. The sample or group of samples that will be used as reference will depend on your experiment system. Be careful, this is a step in which you can make a mistake. Any sample can act as the reference (nontreated control, a time point zero, or healthy individuals), but it must be the same in all the analyses, because the results produced at the end are relative gene expression values.
21. RT-qPCR allows for running many reactions per experiment, so make sure to organize the plates in a specific order to avoid confusions before the run. We usually organize the primers in columns and the different cDNA samples in rows (Fig. 2). Also, the housekeeping gene is located in the first set of columns and the other set of primers are organized in alphabetical order. Assign wells for negative and positive controls for each gene studied. First, we add sterile water as a negative control and we immediately close these wells in order to prevent cross contamination. The following rows are completed with the samples of interest, and the last things to load are the positive controls.
22. Protect the iQ™ SYBR Green SuperMix reagent from light. SYBR Green® and other master mixes of RT-qPCR often contain components such as glycerol that make the solution viscous. This solution should be taken by reverse pipetting. This will prewet the pipette tip and increase the pipette accuracy.
23. Prepare a working dilution of 5 μM primer. The concentration of primers should be optimized in order to avoid nonspecific products or primer-dimer formation. Perform a concentration gradient of primers for each gene, maintaining the other components in the same conditions. The specific products will have an approximate Tm of 80 °C, while curves below Tm 75 °C are characteristic of the primer-dimer or nonspecific products (Fig. 3).
24. Determine the amplification efficiency of the reaction for both the target and reference genes. For this purpose, a serial dilution from pure cDNA to 1/1,000,000 was used to analyze the

efficiency of the PCR reaction for different target genes and housekeeping. The Ct values obtained will allow performing a standard curve for each gene studied. Ideally all the curves should be parallel (same slope) to be analyzed together. PCR efficiencies can be compared by relating the slope of the linear curve of Ct values plotted against the log-dilution. A perfect reaction has a slope of 3.3, which corresponds to an exact two-fold amplification at every cycle of the PCR between dilutions of a factor 10 [28, 29].

The $2^{-\Delta\Delta Ct}$ method assumes that the amplification efficiency for both the target and the reference gene is similar [22].

25. Several factors can affect the variability in the plate, but the most common one is the pipetting technique. As RT-qPCR is an exponential reaction, the effect of small differences in reagents can often have accumulative effect on the Ct value. A minimum error on pipetting may cause some wells to contain different amounts of template, polymerase or primers, leading to variation on the data. Calibrate pipettes periodically; the time recommended for recalibrations every 6–12 months.
26. Use a final concentration of cDNA between 0.001 and 100 ng. Low cDNA concentration or poor quality will decrease levels of amplicons, resulting in high Ct values and variability. All cDNA sample and reactions tubes must be handling cautiously. Try to maintain components capped as much as you can.

Acknowledgments

This project was supported by Estación Experimental Agroindustrial Obispo Colombres (EEAOC) and Consejo Nacional de Investigaciones Científicas y Tecnológicas (CONICET).

References

1. Heid CA, Stevens J, Livak KJ et al (1996) Real time quantitative PCR. Genome Res 6:986–994
2. Galli V, da Silva Messias R, dos Anjos e Silva SD et al (2013) Selection of reliable reference genes for quantitative real-time polymerase chain reaction studies in maize grains. Plant Cell Rep 32:1869–1877
3. Barbau-Piednoir E, Lievens A, Vandermassen E et al (2012) Four new SYBR®Green qPCR screening methods for the detection of Roundup Ready®, LibertyLink®, and CryIAb traits in genetically modified products. Eur Food Res Technol 234:13–23
4. Yi C, Hong Y (2019) Estimating the copy number of transgenes in transformed cotton by real-time quantitative PCR. Methods Mol Biol 1902:137–157
5. Malvick DK (2007) Impullitti AE (2007) detection and quantification of Phialophora gregata in soybean and soil samples with a quantitative, real-time PCR assay. Plant Dis 91:736–742
6. Lin F, Jiang L, Liu Y et al (2014) Genome-wide identification of housekeeping genes in maize. Plant Mol Biol 86:543–554
7. Davidson RM, Hansey CN, Gowda M et al (2011) Utility of RNA sequencing for analysis

of maize reproductive transcriptomes. Plant Genome 4:191–203
8. Gause WC, Adamovicz J (1994) The use of the PCR to quantitate gene expression. PCR Methods Appl 3:S123–S135
9. Higuchi R, Fockler C, Dollinger G et al (1993) Kinetic PCR analysis: real-time monitoring of DNA amplification reactions. Nat Biotechnol 11:1026–1030
10. Page AF, Minocha SC (2005) Analysis of gene expression in transgenic plants. Methods Mol Biol 286:291–312
11. Foy CA, Parkes HC (2001) Emerging homogeneous DNA-based technologies in the clinical laboratory. Clin Chem 47:990–1000
12. Livak KJ, Flood SJA, Marmaro J et al (1995) Oligonucleotides with fluorescent dyes at opposite ends provide a quenched probe system useful for detecting PCR product and nucleic acid hybridization. PCR Methods Appl 4:357–362
13. Bustin SA (2002) Quantification of mRNA using real-time reverse transcription PCR (RT-PCR): trends and problems. J Mol Endocrinol 29:23–39
14. Holland PM, Abramson RD, Watson R et al (1991) Detection of specific polymerase chain reaction product by utilizing the 5′-* 3′ exonuclease activity of Thermus aquaticus DNA polymerase (ofigonucleotide probe/human immunodeficiency virus). Proc Natl Acad Sci U S A 88:7276–7280
15. Tyagi S, Kramer FR (1996) Molecular beacons: probes that fluoresce upon hybridization. Nat Biotechnol 14:303–308
16. Whitcombe D, Theaker J, Guy SP et al (1999) Detection of PCR products using self-probing amplicons and fluorescence. Nat Biotechnol 7:804–807
17. Thelwell N, Millington S, Solinas A et al (2000) Mode of action and application of scorpion primers to mutation detection. Nucleic Acids Res 28:3752–3761
18. Zipper H, Brunner H, Bernhagen J et al (2004) Investigations on DNA intercalation and surface binding by SYBR Green I, its structure determination and methodological implications. Nucleic Acids Res 32:e103
19. Tajadini M, Panjehpour M, Javanmard S (2014) Comparison of SYBR Green and TaqMan methods in quantitative real-time polymerase chain reaction analysis of four adenosine receptor subtypes. Adv Biomed Res 3:85
20. Sambrook J, Russel DW (2001) Molecular cloning: a laboratory manual, 3rd edn. Cold Spring Harbor Laboratory Press, Cold Spring Harbor, NY
21. Pfaffl MW (2004) Quantification strategies in real-time PCR. In: Bustin SA (ed) A–Z of quantitative PCR. International University Line, La Jolla, CA, pp 87–112
22. Livak KJ, Schmittgen TD (2001) Analysis of relative gene expression data using real-time quantitative PCR and the 2-ΔΔCT method. Methods 25:402–408
23. Murphy SP, Simmons CR, Bass HW (2010) Structure and expression of the maize (Zea mays L.) SUN-domain protein gene family: evidence for the existence of two divergent classes of SUN proteins in plants. BMC Plant Biol 10:269
24. Libault M, Thibivilliers S, Bilgin DD et al (2008) Identification of four soybean reference genes for gene expression normalization. Plant Genome 1:44–54
25. Li Z, Hansen JL, Liu Y et al (2004) Using real-time PCR to determine transgene copy number in wheat. Plant Mol Biol Rep 22:179
26. Pfaffl M (2006) Relative quantification. In: Dorak T (ed) Real-time PCR, Quantification strategies in real-time PCR. International University Line, La Jolla, CA, pp 63–82
27. Li Z, Trick HN (2005) Rapid method for high-quality RNA isolation from seed endosperm containing high levels of starch. BioTechniques 38:872. 874, 876
28. Mygind T, Birkelund S, Birkebaek NH et al (2002) Determination of PCR efficiency in chelex-100 purified clinical samples and comparison of real-time quantitative PCR and conventional PCR for detection of Chlamydia pneumoniae. BMC Microbiol 2:17
29. Schmittgen TD, Jiang J, Liu Q et al (2004) A high-throughput method to monitor the expression of microRNA precursors. Nucleic Acids Res 32:e43

Chapter 7

DNA Methylation and Transcriptomic Next-Generation Technologies in Cereal Genomics

Cynthia G. Soto-Cardinault, Fátima Duarte-Aké, Clelia De-la-Peña, and Elsa Góngora-Castillo

Abstract

RNA sequencing (RNA-seq) coupled to DNA methylation strategies enables the detection and characterization of genes which expression levels might be mediated by DNA methylation. Here we describe a bioinformatics protocol to analyze gene expression levels using RNA-seq data that allow us to identify candidate genes to be tested by bisulfite assays. The candidate methylated genes are usually those that are low expressed in a particular condition or developmental stage.

Key words Bioinformatics, Bisulfite technique, Cereals, Genome methylation, Transcriptome expression

1 Introduction

According to the Food and Agriculture Organization of the United Nations (FAO), cereals account for over 60% of food production worldwide. In the last decade, world cereal utilization has increased by 450 million tons. In 2018, the world cereal production was 2611 million tons being rice, wheat and maize the major cereal crops [1]. From this perspective, the development of improved new varieties become very important to ensure food security.

It has been more than a decade since the rice genome was sequenced, the first cereal genome available [2]. Since then, the number of cereal genomes sequenced has increased; to date, the genomes of the following species have been sequenced: rice (*Oryza sativa ssp japonica*) [2–5], maize (*Zea mays*) [6, 7], sorghum (*Sorghum bicolor*) [8, 9], wheat (*Triticum aestivum*) [10], barley (*Hordeum vulgare*) [11], and oat (*Avena sativa*) (in progress) [12]. Different sequencing strategies have been applied to obtain

Cynthia G. Soto-Cardinault and Fátima Duarte-Aké contributed equally to this work.

Luis M. Vaschetto (ed.), *Cereal Genomics: Methods and Protocols*, Methods in Molecular Biology, vol. 2072, https://doi.org/10.1007/978-1-4939-9865-4_7, © Springer Science+Business Media, LLC, part of Springer Nature 2020

the genomic content of the cereals. For instance, roughly, the 389 Mb of rice genome, the ~730 Mb of sorghum genome, and the ~2.3 Gb of maize genome, were sequenced using a BAC-by-BAC approaches based on physical and genetic maps, and standard Sanger methodologies [2, 6, 8, 13].

The advent of next-generation sequencing technologies has revolutionized the field of genomics by sequencing millions of sequences at low cost and high accuracy [14]. The different sequencing platforms (Roche 454 Pyrosequencing, Illumina, Pacific Biosciences (PacBio), Oxoford Nanopore, etc) provide a robust strategy to access to the genomic content of complex cereal species such as Wheat, Barley and Oat. In 2012, the 17 Gb hexaploid bread wheat (*T. aestivum*) genome was sequenced using 454 Pyrosequencing technology [10]. The 5.1 Gb of the Barley (*H. vulgare*) genome were sequenced using a robust strategy including Sanger sequencing, Roche 454 GSFLX Titanium approach and Illumina GAIIx and Hiseq 2000 sequencing platforms [11]. The oat (*O. sativa*) genome whose estimated size is 12.3 Gb, is in process to being sequencing using PacBio approach [12]. Additionally, NGS technologies has allowed to resequence genomes to improve coverage and therefore annotation. For instance, the entirely new assembly of the maize genome (B73 RefGen_v4) is constructed from PacBio Single Molecule Real-Time (SMRT) sequencing at approximately 60-fold coverage and scaffolded with the aid of a high-resolution whole-genome restriction (optical) mapping. The pseudomolecules of maize B73 RefGen_v4 (Accession PRJNA10769) are assembled nearly end-to-end, representing a 52-fold improvement in average contig size relative to the previous reference (B73 RefGen_v3) [15].

The development of genomics and next-generation sequencing (NGS) technologies have enable the discovery of genes and molecular markers associated with important agronomic traits [16, 17]. Many of these traits have been characterized at genome level, however there are many other traits that are poorly understood particularly complex traits that are controlled by many genetic and environmental factors [16]. Thus, sequencing the genome and the identification of all the genes for a cereal crop is only the first step toward understanding the functionality of important agronomic traits.

The gene expression regulation mediated by epigenetic mechanisms such as DNA methylation plays an important role to understand such complex traits. Epigenomics refers to a large-scale study of epigenetics marks on the genome including covalent modifications of histone tails, DNA methylation and small RNA (sRNA) machinery [18, 19]. Studies on the DNA methylation patterns and profiles, has important implications for understanding how certain regions of genomes are expressed under specific developmental contexts, environmental conditions, plant-pathogen interaction, etc. Therefore, a better understanding of the epigenetic modification of genomes and, consequently, its impact on gene expression is likely to have applications in crop improvement [19]. Thus, studies of gene expression regulation have emerged as a key tool for linking DNA sequence methylation to phenotypes. For instance, it has been observed that within several plant species such as mangrove and Agave, it is possible to find populations displaying alternative phenotypes with not genetic variation but a substantial DNA methylation differences [20, 21]. In maize, the accumulation of PEPC and RuBisCO large (rbcL; encoded on chloroplasts) and small (rbcS; nuclear encoded gene) transcripts have been related to DNA methylation at the promoter site induced by light [22]. In rice, a nutrient-specific loss of DNA methylation, which potentially express genes in a context-specific manner, is playing a role in long-term stress adaptation [23].

In this chapter, we will describe a bioinformatics strategy to analyze RNA-seq data, to identify changes in the transcriptome expression profiles that might be associated to DNA methylation, and the analysis for candidate genes by bisulfite conversion method. Briefly, the RNA-seq analysis pipeline described in this chapter addresses (1) a few considerations to sequence a plant transcriptome, (2) preprocessing of raw Illumina data, (3) estimation of transcriptome expression profiles, and (4) visualization of the data. The bisulfite analysis includes (1) DNA extraction, (2) bisulfite conversion, (3) DNA visualization and PCR analysis, and (4) PCR product purification and cloning.

2 Materials

2.1 Maize Genomic Data

The bioinformatics strategy described in this chapter considers the maize genome Gen_ZmB73_v4 as a reference [24] and two RNA-seq libraries from root and shoot maize samples, obtained from the Short Read Archive (SRA) under the accession numbers: SRR822473 and SRR822474, respectively. Briefly, the RNA-seq libraries were sequenced in Single-End (SE) mode using Illumina HiSeq 2000 platform obtaining a total of 11 G for shoot sample and 12 G for root sample [25].

2.2 Software Requirements

The software used in this chapter is under "open source" license [26].

Operating System: Ubuntu 18.04.1 LTS (http://releases.ubuntu.com/18.04/).

Bioinformatics tools: SRAToolkit version 2.8.2 [27], FastQC version 0.11.5 [28], Trimmomatic v.0.38 [29], Bowtie 2 version 2.3.2 [30], HtSeq version 0.10.0 [31], R version 3.5.0 [32].

2.3 Hardware Requirements

The bioinformatics analysis described in this chapter were performed on a Red Hat High Performance Computing (HPC) solution with 19 computing nodes; processors 2 8-cores Intel

Xeon E5–2670; 128 GB/node RAM. And LENOVO-PC with an Intel(R) Core(TM) i7–6700 CPU at 3.40 GHz x86_64 processor; 16G-RAM and Ubuntu 18.04.1 OS.

2.4 Biological Materials

100 mg of maize second leaves from 12-days germinated plantlets were used for DNA methylation analysis. The material was collected from 5 plantlets grown in 50% of relativity humidity, 28 °C and 16/8 photoperiod.

2.5 Reagents, Solutions, Materials, and Instrumentation

1. EZ DNA Methylation Kit (D5001, Zymo Research), which contains the CT-Conversion Reagent, M-Dilution Buffer, M-Binding Buffer, M-Wash Buffer, M-Desulfonation Buffer, M-Elution Buffer, and Zymo-Spin™ IC Columns.
2. Qubit dsDNA kit (Invitrogen) (optional).
3. Wizard® SV Gel and PCR cleanup system.
4. Pure Link Plasmid Mini Prep kit (Invitrogen).
5. PCR Cloning kit Zero Blunt TOPO for sequencing.
6. RNase A (20 mg/mL stock, Invitrogen).
7. PCR4 Blunt-TOPO vector (Thermo Fisher).
8. Ampicillin (A939, Sigma-Aldrich).
9. TE buffer: 10 mL of 1 M Tris–HCl pH 8.0, 2 mL of 0.5 M EDTA solution to 1000 mL with distilled H_2O.
10. CTAB buffer (100 mM Tris–HCl pH 8.0; 20 mM EDTA; 1.4 M NaCl; 2% CTAB; 2% PVP).
11. Chloroform: Isoamyl-OH (24:1).
12. Agar LB media: 8 gL^{-1} agar (A7002, Sigma-Aldrich), 20 gL^{-1} LB Broth Lennox (L3022, Sigma-Aldrich).
13. Hyperladder IV size marker (Bioline).
14. Loading buffer.
15. Liquid nitrogen.
16. Mortar and pestle.
17. Micropipettes and tips (2–20 μL, 20–200 μL, and 100–1000 μL).
18. Falcon tubes (50 mL).
19. Centrifuge tubes (50 mL).
20. Eppendorf centrifuge tubes (1.5 and 2 mL).
21. Refrigerated centrifuge.
22. Vortex.
23. Rotating mixer.
24. Freezer (−80 °C).
25. Cold room or refrigerator (4–8 °C).

26. Thermocycler.
27. Agarose electrophoresis equipment.
28. UV transilluminator.
29. NanoDrop (ND-1000 spectrophotometer-Thermo).
30. Qubit lector (Invitrogen) (optional).

3 Methods

3.1 Transcriptome Sequencing Considerations

To measure transcriptome expression a sequencing platform that provides deep sequencing is recommended, such as Illumina. Usually, the sequencing cost set the limits to the amount of sequences to generate and, consequently, the biological outcomes. However, there are a few issues to be considered [33, 34]:

1. Single-end (SE) or Paired-end (PE) reads. The use of paired-end reads rather than single end reads will significantly improve a de novo transcriptome assembly, however, paired end reads are more expensive and time-consuming to perform than single-end reads. In paired-end mode (PE), the reading starts at on end, finishes this direction, and then starts another round of reading from the opposite end of the fragment.
2. The read length. It is possible to specify the number of base pairs for the read length. Longer reads can provide more reliable information and is a major driver in the robustness of a de novo transcriptome assembly, however it is usually more expensive to generate longer reads.
3. Coverage. Determining coverage for RNA sequencing is complicated due to transcripts are expressed at different levels. This means that more reads will be sequenced from highly expressed genes, and few sequences will be captured for low expressed genes. Several reviews articles highlight the considerations for coverage estimation and the reader is referred to these for an in-depth discussion of the coverage for RNA-seq studies [35, 36].

3.2 Quality Assessment of Sequences Using FastQC

The majority of high throughput sequencing platforms run under similar principle. First, the chemistry reactions requires to add platform specific adapter sequences to the ends of each DNA sequence to provide anchorage and priming points for the sequencing reaction. Second, sequencing progresses by running a chemistry cycle which adds a tagged nucleotide to extend the sequence. As the number of sequencing cycles performed is increased the average quality of the base calls falls. The rate at which this fall happens will vary according to the type of sequencer used, the version of the sequencing chemistry and the nature of the library being sequenced [37].

Thus, to identify low quality reads and adapter sequences, prior any downstream analysis, a QC checking of the sequences is necessary.

FastQC is a quality control application for NGS data [28]. The output is an HTML file containing a series of QC analyses. Each analysis is flagged as pass, warning or fail. The HTML file can be visualized using an internet browser such as Chrome, Firefox, etc. Briefly, the QC analyses include the following measurements: (1) Per base sequence quality, (2) Per sequence quality scores, (3) Per base sequence content, (4) Per base GC content, (5) Per sequence GC content, (6) Per base N content, (7) Sequence length distribution, (8) Sequence duplication levels, (9) overrepresented sequences, and (10) Kmer content.

Box 1: FastQC command line. The *–extract* flag show the output file in unzipped format; the *–outdir* flag points out to the output directory; the *–f* flag indicates the format of in the input file

A typical FastQC command in the Linux shell follows the next syntax:

```
$ fastqc --extract --outdir=</path/to/my/output_
directory> -f fastq <raw_sequences_file1.fastq> ...
<raw_sequences_fileN.fastq>
```

The following command line was applied to the Illumina data used for the purpose of this chapter:

```
$ fastqc --extract --outdir=~/fastqc_raw_data/ -f
fastq ~/SRR822473_root.fastq ~/SRR822474_shoot.
fastq
```

Per base sequence quality analysis is represented in a BoxWhisker plot, this section of the report is probably the one that receives most attention. A boxplot is produced for every base-position in the read shown on the *x*-axis. The central line and yellow box represent the median and interquartile range in the usual manner (Fig. 1). The quality for each base is shown on the *y*-axis based on Phred quality scores. The overrepresented sequences analysis will show if the dataset is contaminated with adapter sequences. The reader is referred to the manual for more information about the FASTQC output [28] (*see* **Notes 1** and **2**).

3.3 Removing Low Quality and Adapter Sequences Using Trimmomatic

Trimmomatic is a command line application to remove adapters, trim and crop Illumina sequences in fastq format [29] in which the parameters are specified by the user (Box 2). Since the sequence of the adapters used for a given library is known ahead of time, it is possible to explicitly screen for that sequence within the library to identify which reads contain adapter. Adapter trimming is the standard approach to remove the adapter sequence. A search for the adapter sequence is performed for each read, and the read is truncated to the point where the adapter is found to start.

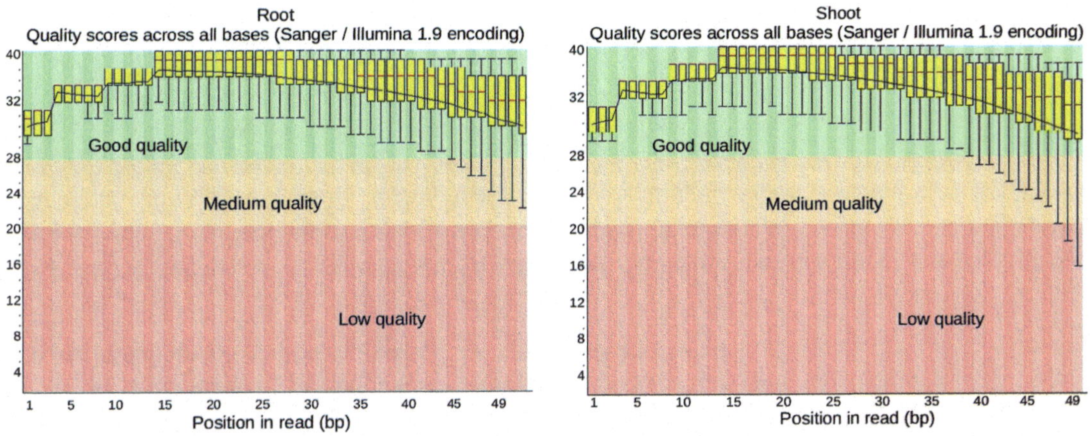

Fig. 1 Reads quality overview for root and shoot RNA-seq data. The BoxWhisker plots show the quality scores (*y*-axis) for all the bases at each position of the read (*x*-axis). The background of the graph is divided in three colors: (1) green for the good quality calls, (2) yellow for reasonable quality calls, and (3) red for poor quality calls

Box 2: Trimmomatic command line. The *ILLUMINACLIP* argument points out to the file TruSeq2 (used in Illumina GAII platform) and TruSeq3 (used by Illumina HiSeq and MiSeq platforms) that contains a list of adapter sequences to be trimmed. *LEADING* and *TRAILING* arguments remove low quality or N bases below the quality number specified by user. For Illumina reads a minimum quality of 20 (based on Phred-score) is recommended. *SLIDINGWINDOW* argument is used for scanning the reads with a 4-base sliding window and cutting when the average quality per base drops below 20

The basic command line for SE data follows the next syntax:

```
$ java -jar trimmomatic-0.38.jar SE -phred33 input.fq.gz output.fq.gz ILLUMINACLIP:TruSeq3-SE.fa:2:30:10 LEADING:20 TRAILING:20 SLIDINGWINDOW:4:15 MINLEN:50
```

The following command line was used for the purpose of this chapter:

```
$ java -jar trimmomatic-0.38.jar SE -phred33 SRR822473_root.fastq out_root.fastq ILLUMINACLIP:TruSeq3-SE.fa:2:30:10 LEADING:20 TRAILING:20 SLIDINGWINDOW:4:20 MINLEN:30
```

The reader is referred to Trimmomatic manual for more information (*see* **Notes 2** and **3**).

3.4 Genome Alignment Using Bowtie2

The sequences aligners use "sequence similarity" strategies to infer how and where the reads are located within a sequence reference (i.e., a genome). Bowtie 2 is an ultrafast and memory-efficient command line tool for aligning sequences to a long reference [30, 38]. The first step to align sequences to the reference genome is to build a genome-based index (Box 3).

Box 3: Bowtie-build command line. The *-f* argument specifies the reference input file is in FASTA format; followed by genome sequence FASTA file and the base-name of the index file. For the purpose of this chapter the maize genome v4 was used as reference (Gen_ZmB73_v4; https://www.maizegdb.org/)

The basic command line is:

```
$ bowtie2-build [options]* <reference_in> <bt2_base>
```

The following command line was used to build a maize genome (v4) index:

```
$ bowtie2-build -f /path/to/file/Zea_mays.B73_RefGen_v4.dna.toplevel.fa /path/to/file/bowtieIdx
```

By default, bowtie2-build writes six small index files; in this example the files were named as bowtieIdx.1.bt2, bowtieIdx.2.bt2, bowtieIdx.3.bt2, bowtieIdx.4.bt2, bowtieIdx.rev.1.bt2, and bowtieIdx.rev.2.bt2.

Bowtie 2 performs an end-to-end alignment by default, which means that searches for alignments involving all the bases within a read. However, bowtie 2 also supports local alignments which does not require reads to align entirely. For local alignments, bowtie 2 might "trim" some read characters from one of both ends of the alignment to maximize the alignment score. The best possible alignment score in end-to-end mode is 0, it means there are no difference between the read and the reference. A score is calculated by subtracting penalties for each difference found (mismatch, gap, etc.). Conversely, in local alignment mode, the score is calculated adding bonuses for each match.

Box 4: Bowtie2 command line. The *-x* argument points out to index files. Single-end (SE) reads are specified with *-U* argument

The basic command line is:

```
$ bowtie2 -x <bowtieIdx_basename> -U <input_file_name> -S <output_file.SAM>
```

The command line to align high quality reads from root and shoot RNA-seq libraries to the maize reference genome, were:

```
$ bowtie2 -x /path/to/file/Corn_genome_Idx/bowtieIdx -U /path/to/file/out_root.fastq -S /path/to/file/bowtie_out_root.sam

$ bowtie2 -x /path/to/file/Corn_genome_Idx/bowtieIdx -U /path/to/file/out_shoot.fastq -S /path/to/file/bowtie_out_shoot.sam
```

The bowtie2 output file is a set of alignments in SAM format. The index is specified only by the file base-name (i.e., bowtieIdx), the file extension (0.1.bt2, 0.2.bt2, etc.) are not included (Box 4).

The output file is a set of alignments in SAM format, in which each line is a collection of at least 11 mandatory fields separated by tabs [39]. Thus, from left to right, the fields are (1) Name of read, (2) Sum of all applicable flags, (3) Name of reference sequence, (4) 1-based offset into the forward reference strand where leftmost character of the alignment occurs, (5) Mapping quality, (6) CIGAR string representation of alignment, (7) Name of reference sequence where mate's alignment occurs, (8) 1-based offset into the forward reference strand where leftmost character of the mate's alignment occurs, (9) Inferred fragment length, (10) Read sequence, and (11) ASCII-encoded read qualities, and (12) subsequent optional fields can be specified by the user.

At the end of the run, bowtie2 provides a summary of the alignment statistics and the overall alignment rate (Table 1).

The reader is referred to Bowtie 2 manual and SAMtools format guide for more details (*see* **Notes 2**, **4**, and **5**).

3.5 Gene Expression Quantification Using HTSeq

HTSeq is a python-based command line application that simplifies working with data associated to genomic coordinates, for instance, values attributed to genomic positions or genomic intervals (genomic features such as exons or genes). One of the HTSeq tools is *htseq-count* that process RNA-seq alignments for expression analysis. The *htseq-count* application takes one or more alignment files in SAM/BAM format and calculates the number of mapped reads for each feature according to a General Feature Format (GFF) file [31] (Box 5).

Box 5: Htseq-count command line. The most relevant arguments are: *--format*, describes the alignment file format (SAM or BAM); *--stranded*, specifies whether the data is from a strand-specific assay (default: yes); *--type*, that points out to the feature type located in the third column of the GFF file; and *–idattr*, which is the GFF attribute to be used as feature ID

A typical command line to execute *htseq-count* is:

```
$htseq-count [options] <alignment_file> <gff_file>
> output.file.txt
```

The following command line was used to quantify maize root RNA-seq reads aligned to maize reference genome (*see* **Note 6**):

```
$ htseq-count --mode=union --nonunique none
--stranded=no --format=sam --idattr=gene_id
--type=gene /path/to/file/bowtie_out_root.sam /
path/to/file/Zea_mays.B73_RefGen_v4.42.gff3 > /
path/to/file/gene_root_count.txt
```

Assigning a read to a feature (gene, exon, etc.) is not a trivial task, *htseq-count* operates using three different modes to handle reads that overlaps several genes or features through the *–m* (mode) argument. The "union" mode is set up by default, and it is recommended for most use cases, in which the read is assigned to

Table 1
Summary of the alignment statistics and the overall alignment rate

	Root	Shoot
Total reads	44,756,082	51,002,585
Unpaired reads	44,756,082 (100%)	51,002,585 (100.00%)
Aligned 0 times	6,193,636 (13.84%)	6,747,888 (13.23%)
Alligned exactly 1 time	25,056,340 (55.98%)	28,419,416 (55.72%)
Aligned >1 times	13,506,106 (30.18%)	15,835,281 (31.05%)
Overall alignment rate	86.16%	86.77%

Table 2
Example of an expression matrix of maize root and shoot transcriptome abundances expressed in TPM values

	Raw counts from *htseq-count*		Normalization factors				Expression values in TPM	
Gene name	Score root	Score shoot	Gene length	Gene length (Kb)	RPK root	RPK shoot	TPM root	TPM shoot
Zm00001d031168	70,306	28,943	1119	1.119	62,829	25,865	13,099	3775
Zm00001d048611	28,298	16,328	772	0.772	36,655	21,150	7642	3086
Zm00001d027652	68,114	31,545	2055	2.055	33,145	15,350	6910	2240
Zm00001d029558	14,247	2	539	0.539	26,432	4	5511	1
Total RPK[a]:					4,796,529	6,852,518		
Per million factor:					5	7		

[a]The table shows only four genes. The total RPK value includes the sum of all the 39,592 genes in the maize genome

a feature if a given read overlap entirely, partially (more than 50% of the read length) or it is separated by an intron. The output file is a tabular file containing in the first column the attribute used as feature ID, in this case the "gene_id", and a second column containing the raw counts for each selected feature in the genome (i.e., gene) (Table 2; *see* **Note 6**). Reads classified as "__ambiguous" are those which could have been assigned to more than one feature and hence were not counted for any of the features; reads classified as "__no_feature" are those which could not be assigned to any feature. These "special counters" are located at the bottom of the output file. The reader is referred to HTSeq website and manual for more detailed information about this application (*see* **Note 2**).

3.6 Data Normalization Using Transcripts Per Kilobase Million (TPM)

To compare RNA levels across different samples a data normalization is required. Various normalization approaches have been proposed and it is a key step of data processing as it has a profound effect on the results of the analysis. The aim of normalization is to handle sequencing depth variation and gene length. The most frequently reported RNA-seq expression values are: RPM (Reads Per Million mapped reads), RPKM or FPKM (Reads or Fragments Per Kilobase of exon model per Million mapped reads) and TPM (Transcripts Per Kilobase Million) [40].

There are several excellent reviews on RNA-seq data normalization in which the limitations for each method are deeply discussed. The reader is referred to these papers for additional details [36, 41–44] (*see* **Note 2**).

Transcripts Per Kilobase Million (TPM) method has been reported to effectively normalize for the differences in composition of the transcripts rather than simply dividing the number of reads in the library [36, 43]. Thus, this method is considered more effective to compare samples of different origins and composition (although it is not free from bias). In the TPM normalization method, gene length is normalized first and sequencing depth second. Using this method the summation of all TPMs for each sample are the same, which makes easier to compare the proportion of mapped reads in each sample [44]. Consequently, an expression matrix containing TPM expression values is created for downstream analysis (Table 2; *see* **Note 7**).

To calculate TPM [44]:

1. To obtain "reads per kilobase (RPK)", divide the reads counts by gene length expressed in kilobases (*see* **Note 7**; Table 2).
2. To obtain "per million" factor, calculate the sum of all RPK values in a sample and divided by million (Total RPK/1,000,000) (Table 2).
3. To obtain TPM, divide each RPK value by "per million" factor (Table 2).

3.7 Gene Expression Visualization Using Heatmaps

A heatmap is a graphical representation of data where individual values contained in a matrix are represented as colors [45]. A heatmap allows to simultaneously visualize clusters of samples and features and can be created using a function of R [46]. A few considerations to manage the data for a heatmap plotting: (1) Use normalize data; (2) Transforming the data to a logarithmic scale is useful to reduce the large scale when dataset covers a broad range of values (*see* **Note 8**); (3) Normalized and transformed expression values can be storage in a "Comma Separated Value (CVS)" file format to be read by R (Box 6). For the purpose of this chapter, a subset of maize root and shoot genes were selected and plotted as heatmap (Fig. 2).

Box 6: The basic pipeline to create a heatmap for Log2 transformed TPM expression data of maize root and shoot. The most relevant arguments for heatmap function are described in **Note 9**

```
library(gplots)
   library(RColorBrewer)
   library(heatmap.plus)
   # Set of 50 normalized values y TPM and trans-
formed with Log2
   data   <-   read.csv("~/Xtromes-TPM-Htseq-Log2.
csv")
   # Getting rows names to transform frame values
into matrix values
   rnames <- data[,1]
   mat_data <- data.matrix(data[,2:ncol(data)])
   # Assign the column names, previously saved as
"rnames"
   rownames(mat_data) <- rnames
   # Different color names in R can be found at
   #   http://www.stat.columbia.edu/~tzheng/files/
Rcolor.pdf
   coul = colorRampPalette(brewer.pal(8, "PiYG"))
(25)
   # Saving the heatmap as PNG file.
   # Creating a 5x5 inch image with 300 pixels per
inch.
   # [1500 pixels] / [300 pixels/inch] = 5 inches.
   # Font size of 8 pt.
   png("~/heatmap-X.png",      width      =5*300,
height=5*300, res=300, pointsize=8)
   # Plotting the heatmap
   heatmap.2(mat_data,          cellnote=mat_data,
main="Transcriptome      Expression      Profile",
notecol="black",density.info="none",trace="none",
cexRow=0.75, cexCol=0.75, margins=c(3,6), col=coul,
dendrogram="row",Colv="NA")
   # Closing device to access to the png object
   dev.off()
```

Heatmaps plots are an excellent tool for expression data visualization. The reader is referred to great online tools and tutorials for more in-depth information about heatmaps (*see* **Note 2**).

3.8 Bisulfite Analysis

Once the transcriptome expression profile revealed the candidate genes (Fig. 2), the next step is validation of the expression level related to DNA methylation along the complete sequence of the gene. For this propose, bisulfite analysis allows to identify the methylated status and the context of methylation of a gene [47]. This technique is widely used to analyze the context of methylation and to understand the mechanisms by which the epigenetic regulation

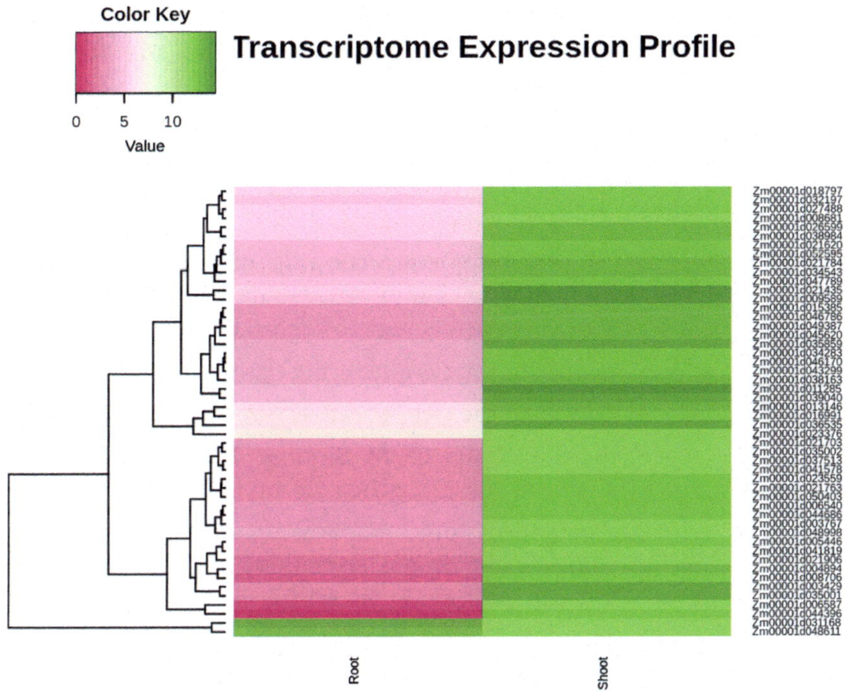

Fig. 2 Maize transcriptome expression profiles. A subset of maize genes were selected to plot the shoot and root transcriptomes expression profiles. The raw counts derived from mapped reads to the maize (v4) genome were normalized by transcripts per kilobase million (TPM). The TPM expression values were transformed as Log2(TPM+1) and plotted using heatmap.2 from R

of a specific gene drives the expression level. The main steps to analyze the content of methylated DNA are described as follow:

3.8.1 DNA Extraction

1. Total DNA is extracted from 100 mg of fresh tissue, grinded in a mortar with liquid nitrogen and homogenate with 1 mL of preheated CTAB buffer at 65 °C.
2. The homogenate is transferred to a 2 mL Eppendorf tube.
3. The sample is incubated for 30 min at 65 °C and 10 μL of RNase A is added.
4. The sample is incubated for 30 min at 37 °C.
5. Add 900 μL CI:Isoamyl-OH, mix gently, centrifuge 3 min at 13,200 × *g* and recover 800 μL from the upper phase to a new Eppendorf tube.
6. Add 800 μL CI, mix gently and centrifuge 3 min at 13,200 × *g*.
7. Take the upper phase to a new Eppendorf and add 600 μL of isopropanol and mix gently.
8. Incubate the sample for 1 h on ice, centrifuge 20 min at 13,800 × *g* at 4 °C and wash twice with 600 μL of 80% cold ethanol.

9. Dry the pellet and dissolve it in 50 μL of TE, 30 min at 55 °C.
10. Check DNA quality and quantity and run a gel: 0.8% at 80 V. Verify the DNA quantity.

3.8.2 Bisulfite Conversion

1. Use 500 ng of DNA add 5 μL of M-Dilution Buffer (provided in EZ DNA Methylation Kit) to the DNA sample and adjust the total volume to 50 μL with water (*see* **Note 10**).
2. Mix the sample pipetting up and down. Incubate the samples at 37 °C for 15 min. After that add 100 μL of the prepared CT Conversion Reagent to each sample and mix very well.
3. Incubate the sample in the dark at 50 °C for 12–16 h.
4. Incubate the sample on ice for 10 min (*see* **Note 11**).
5. Add 400 μL of M-Binding Buffer to a Zymo-Spin™ IC Column and place the column into a provided collection tube.
6. Load the sample that was incubated on ice into the Zymo-Spin™ IC Column containing the M-Binding Buffer. Close the cap and mix by inverting the column 10 times.
7. Centrifuge at 10,000 × *g* for 1 min. Discard the flow-through.
8. Add 100 μL of M-Wash Buffer to the column and centrifuge at 10,000 × *g* for a min.
9. Add 200 μL of M-Desulfonation Buffer to the column and let stand for 20 min at 25 °C. After the incubation, centrifuge the sample at 10,000 × *g*.
10. Add 200 μL of M-Wash Buffer to the column and centrifuge at 10,000 × *g* for 30 s.
11. Add another 200 μL of M-Wash Buffer and centrifuge for an additional 30 s.
12. Place the column into a 1.5 mL microcentrifuge tube and add 10 μL of M-Elution Buffer directly to the column matrix.
13. Centrifuge the sample for 30 s at full speed to elute the DNA. The DNA is ready for immediate analysis or can be stored at or below −20 °C for later use.

3.8.3 DNA Visualization and PCR Analysis

1. 100 μg of the converted DNA is analyzed in 2% agarose gel by electrophoresis to measure the recovery and fragmentation of DNA.
2. After the electrophoresis, chill on ice the gel for 5–10 min (*see* **Note 12**). After this, the expected sweep between 1500 and 100 bp can be visualized in a gel-digitizer.
3. For PCR analysis, mix 200 ng DNA templates, 1 μM forward primer, 1 μM reverse primer, 0.2 μL dNTPs, 0.2 μL high-fidelity Taq Polymerase and 4 μL of Taq buffer in a total volume of 20 μL (*see* **Note 13**).

4. The PCR program consist in 35 cycles of 20 s at 95 °C of denaturation, annealing for 30 s at 55–60 °C and an extension period of 60 s at 62 °C and a final extension of 10 min at 62 °C.
5. PCR products have to be examined on 2% agarose gel, by loading 5 μL PCR product and 5 μL loading buffer. Use a Hyperladder IV size marker to determine the molecular weight of the products.

3.8.4 PCR Product Purification and Cloning

1. Purify the amplified products using the Wizard® SV Gel and PCR Clean-Up System according to the manufacturer's protocol.
2. Clone the amplified PCR products into PCR 4 Blunt-TOPO vector using PCR Cloning kit Zero Blunt TOPO for Sequencing, as per the manufacturer's instructions and transformed into *Escherichia coli* DH5α competent cells.
3. Culture bacteria on LB agar selective media containing 100 mg mL^{-1} ampicillin and the transformed colonies can be screened by PCR. Then, plasmids are purified with Pure Link Plasmid Mini Prep kit, and 25–50 ng of plasmid is ready to sequencing.

4 Notes

1. For Paired-End reads, the quality assessment and preprocessing of the reads need to be applied to each file in the pair.
2. Useful Links.
   - FASTQC manual:

     https://biof-edu.colorado.edu/videos/dowell-short-read-class/day-4/fastqc-manual
   - Trimmomatic manual:

     http://www.usadellab.org/cms/uploads/supplementary/Trimmomatic/TrimmomaticManual_V0.32.pdf
   - Bowtie 2:

     http://bowtie-bio.sourceforge.net/bowtie2/manual.shtml
   - SAMtools:

     http://samtools.sourceforge.net/
   - HTSeq:

https://htseq.readthedocs.io/en/release_0.11.1/

https://htseq.readthedocs.io/en/release_0.11.1/count.html

https://htseq.readthedocs.io/en/master/count.html

- GFF file format:

  https://www.ensembl.org/info/website/upload/gff.html

- RPKM, FPKM and TPM:

  https://www.rna-seqblog.com/rpkm-fpkm-and-tpm-clearly-explained/

- Heatmaps tutorials:

  https://jcoliver.github.io/learn-r/006-heatmaps.html

  https://www.rdocumentation.org/packages/stats/versions/3.5.3/topics/heatmap

  https://sebastianraschka.com/Articles/heatmaps_in_r.html

  https://www.statisticshowto.datasciencecentral.com/probability-and-statistics/z-score/

3. Preprocessing steps could be time consuming depending on the file size. It is recommended to perform the preprocessing of the sequences based on the user needs and not running the applications by default options.
4. Bowtie 2 (.bt2) index format is different from Bowtie 1 (.ebwt) format, and they are not compatible with each other.
5. Bowtie 2 is recommended for reads longer than 50 bp and there is no upper limit on read length. For relatively short reads (less than 50 bp) bowtie 1 is faster and more sensitive and the upper limit of read length is about 1000 bp.

   When aligning PE reads the file with forward reads should be specified using the –1 argument and the reverse reads should be specified using –2 argument. The basic command line to run PE mode is:

```
$ bowtie2 -x /path/to/bowtie_index -1 /path/to/fastq_1.fq -2 /path/to/fastq_2.fq -S /path/to/output_file.sam
```

6. The htseq-count output file is a text tabular file containing usually two columns, the first column is the feature (in this example "gene_id" was used as a feature) and the second column shows the raw counts for each feature. One output file per sample is created. An example for root and shoot raw counts file, respectively, is shown:

Root counts:

Zm00001d031168	70,306
Zm00001d048611	28,298
Zm00001d027652	68,114
Zm00001d029558	14,247

Shoot counts:

Zm00001d031168	28,943
Zm00001d048611	16,328
Zm00001d027652	31,545
Zm00001d029558	2

The next step is create a united matrix (table) file, containing the raw counts for all the samples. The first column should display the feature ("gene_id"). Second, third, etc., columns should display the raw counts for each gene of each sample.

7. The first step to obtain TPM expression values is to calculate the length for each gene in the genome. This can be done by processing the GFF file. Due to the GFF file size and the versatility of Linux, the Linux shell is the best option to process this file. The *egrep* and *cut* commands can be applied as a pipeline to obtain the information for each gene. The *egrep* command would select all the rows containing the word "gene"; *cut* command, with the *-f* flag, would cut the selected columns. In this example the column 4 contains the starting coordinate of the gene, the column 5 contains the ending coordinate of the gene and column 9 contains the gene name.

   The command line to parse the maize GFF file is:

```
$ egrep "gene" path/to/Zea_mays.V4.gff3 | cut
-f 4,5,9 > /path/to/maize_gene_coords_file.txt
```

   To sort the columns in the output file use the *sort* command. The *-k* flag indicates to sort by specific column. The *awk* command prints sorted columns by gene name, starting coordinate and ending coordinate separated by comma. The output file will contain the gene name, the first and last coordinate for each gene:

```
$ sort -k 3 /path/to/maize_gene_coords_file.
txt | awk '{print $3 "," $1 "," $2}' > /path/
to/maize_gene_coords_sorted_file.txt
```

   A second step needs to be performed to obtain the length for each gene by calculating the difference between the last and first coordinate.

   The maize genome file and GFF file for maize reference genome (v4) was obtained from ftp://ftp.gramene.org/pub/gramene/release-60/fasta/zea_mays/dna/

8. The *scale* attribute in a heatmap indicates if the values should be centered and scaled. The default option is "row" and automatically transform the data into a Z-score metric [−1α, −0, +1α]. The Z-score is a measure of distance in standard deviation from the plate mean. The reader is referred to several great tutorials and on line tools for more information about Z-score calculations (*see* **Note 2**).

   On the other hand, using a log-transformed expression values models proportional changes in the gene expression rather than additive changes. Transforming the values to Log2(TPM+1) will transform zero counts to 1 and log2 will compress the scale keeping the proportional changes in gene expression (Fig. 2).

9. All the arguments for a heatmap.2 can be defined by the user. In Fig. 2 the arguments used were: *cellnote* (matrix of character strings which will be placed within each color cell); *notecol* (specifies the color for *cellnote*); *main* (defaults to none); *density.info* (indicates whether to superimpose a "histogram," a "density" plot or "none" on the color key); *trace* (indicates whether a solid trace line should be drawn across "rows," "columns," "both," or "none." Defaults to "column"); *cexCol, cexRow* (a positive number for the row or column axis labeling); *margins* (a numeric vector of length 2 containing the margins for column and row names, respectively); *col* (color used for the image); dendrogram (indicates whether to draw "none," "row," "column," or "both" dendrograms. Defaults to "both"); *Colv* (determines how the column dendrogram should be reordered).

   For more information about heatmap.2 arguments: https://www.rdocumentation.org/packages/gplots/versions/3.0.1.1/topics/heatmap.2

10. Another options for Bisulfite conversion kits are: MethylDetector Bisulfite Modification Kit (Active Motif), Methyl SEQr Bisulfite Conversion Kit (Applied Biosystems), Methyl Easy Kit (Diagenode), MethylCode Bisulfite Conversion Kit (Invitrogen).

11. Samples may be storage at 4 °C up to 20 h.

12. Cool down the gel for 5–10 min, allowing for the insertion of the ethidium bromide to the DNA to make it visible.

13. The oligos design is a key step to obtain the gene fragment to analyze. The primer size is important for bisulfite PCR. The primers are usually between 26–30 bases and the resulted amplicon should be relatively short, between 150 and 300 bp. During the primer design is important to consider that primers should not contain CpG sites. It is also important to note that only one strand of the bisulfite-converted template will be amplified by any given primer set. The free-online tool

Bisulfite Primer Seeker by Zymo Research or Kismeth program is recommended for primer design [48].

To ensure optimal amplification of the specific target, test the annealing temperature gradient for every new primer set. Additionally, 35 to 40 cycles are recommended for amplification. The cycles start at 95 °C for nonspecific amplification. The optimal annealing temperature is 55–60 °C.

Acknowledgments

The authors work was supported by two grants received from the National Council for Science and Technology (CB2016-285898, CB2016-286368 and INFR-2016-01-269833) and Cátedras Marcos Moshinsky 2017.

References

1. FAO (2019) Cereal supply and demand brief. World food situation. www.fao.org/world-foodsituation/csdb/en. Accessed 24 Feb 2019
2. International Rice Genome Sequencing Project (2005) The map-based sequence of the rice genome. Nature 436:793–800
3. Ohyanagi H (2006) The Rice Annotation Project Database (RAP-DB): hub for Oryza sativa ssp. japonica genome information. Nucleic Acids Res 34:D741–D744
4. Ouyang S, Zhu W, Hamilton J et al (2007) The TIGR Rice genome annotation resource: improvements and new features. Nucleic Acids Res 35:D883–D887
5. Kawahara Y, de la Bastide M, Hamilton JP et al (2013) Improvement of the Oryza sativa Nipponbare reference genome using next generation sequence and optical map data. Rice 6:4
6. Schnable PS, Ware D, Fulton RS et al (2009) The B73 maize genome: complexity, diversity, and dynamics. Science 326:1112–1115
7. Vielle-Calzada J-P, Martínez de la Vega O, Hernández-Guzmán G et al (2009) The Palomero genome suggests metal effects on domestication. Science 326:1078
8. Paterson AH, Bowers JE, Bruggmann R et al (2009) The Sorghum bicolor genome and the diversification of grasses. Nature 457:551
9. McCormick RF, Truong SK, Sreedasyam A et al (2018) The *Sorghum bicolor* reference genome: improved assembly, gene annotations, a transcriptome atlas, and signatures of genome organization. Plant J 93:338–354
10. Brenchley R, Spannagl M, Pfeifer M et al (2012) Analysis of the bread wheat genome using whole-genome shotgun sequencing. Nature 491:705–710
11. The International Barley Genome Sequencing Consortium (2012) A physical, genetic and functional sequence assembly of the barley genome. Nature 491:711–716
12. Schlueter J (2019) The oat genome project. AVENA GENOME. avenagenome.org. Accessed 20 Feb 2019
13. Gupta PK, Varshney RK (eds) (2005) Cereal genomics. Springer, Dordrecht
14. Shendure J, Balasubramanian S, Church GM et al (2017) DNA sequencing at 40: past, present and future. Nature 550:345–353
15. Jiao Y, Peluso P, Shi J et al (2017) Improved maize reference genome with single-molecule technologies. Nature 546:524
16. Edwards D, Batley J (2010) Plant genome sequencing: applications for crop improvement: plant genome sequencing: applications for crop improvement. Plant Biotechnol J 8:2–9
17. Visendi P, Batley J, Edwards D (2013) Next generation characterisation of cereal genomes for marker discovery. Biology 2:1357–1377
18. Rival A, Beulé T, Aberlenc Bertossi F et al (2010) Plant epigenetics: from genomes to epigenomes. Not Bot Hortic Agrobot Cluj-Napoca 38:09–15
19. Edwards MA, Henry RJ (2011) DNA sequencing methods contributing to new directions in cereal research. J Cereal Sci 54:395–400
20. Duarte-Aké F, Castillo-Castro E, Pool FB et al (2016) Physiological differences and changes in global DNA methylation levels in Agave angustifolia Haw. albino variant somaclones

during the micropropagation process. Plant Cell Rep 35:2489–2502
21. Lira-Medeiros CF, Parisod C, Fernandes RA et al (2010) Epigenetic variation in mangrove plants occurring in contrasting natural environment. PLoS One 5:e10326
22. Langdale JA, Taylor WC, Nelson T (1991) Cell-specific accumulation of maize phosphoenolpyruvate carboxylase is correlated with demethylation at a specific site >3 kb upstream of the gene. Mol Gen Genet 225:49–55
23. Mager S, Schönberger B, Ludewig U (2018) The transcriptome of zinc deficient maize roots and its relationship to DNA methylation loss. BMC Plant Biol 18:372
24. Portwood JL, Woodhouse MR, Cannon EK et al (2019) MaizeGDB 2018: the maize multi-genome genetics and genomics database. Nucleic Acids Res 47:D1146–D1154
25. Diepenbrock CH, Kandianis CB, Lipka AE et al (2017) Novel loci underlie natural variation in vitamin E levels in maize grain. Plant Cell 29:2374
26. Atkinson L (2019) Open source initiative. March 2019 license-discuss summary. opensource.org. Accessed 16 Apr 2019
27. SRA Toolkit Development Team SRA-Tools. SRA Toolkit Documentation. ncbi.github.io/sra-tools/. Accessed 14 Feb 2019
28. Andrews S (2010) FastQC. A quality control tool for high throughput sequence data. bioinformatics.babraham.ac.uk/projects/fastqc. Accessed 10 Apr 2018
29. Bolger AM, Lohse M, Usadel B (2014) Trimmomatic: a flexible trimmer for Illumina sequence data. Bioinformatics 30:2114–2120
30. Langmead B, Salzberg SL (2012) Fast gapped-read alignment with bowtie 2. Nat Methods 9:357–359
31. Anders S, Pyl PT, Huber W (2015) HTSeq--a python framework to work with high-throughput sequencing data. Bioinformatics 31:166–169
32. R Core Team (2013) R: A language and environment for statistical computing. https://www.r-project.org/. Accessed 2 Feb 2019
33. Góngora-Castillo E, Buell CR (2013) Bioinformatics challenges in de novo transcriptome assembly using short read sequences in the absence of a reference genome sequence. Nat Prod Rep 30:490–500
34. Góngora-Castillo E, Fedewa G, Yeo Y et al (2012) Genomic approaches for interrogating the biochemistry of medicinal plant species. Methods Enzymol 517:139–159
35. Sims D, Sudbery I, Ilott NE et al (2014) Sequencing depth and coverage: key considerations in genomic analyses. Nat Rev Genet 15:121–132
36. Conesa A, Madrigal P, Tarazona S et al (2016) A survey of best practices for RNA-seq data analysis. Genome Biol 17:13. https://doi.org/10.1186/s13059-016-0881-8
37. Andrews S (2016) Loss of base call accuracy with increasing sequencing cycles. https://sequencing.qcfail.com/. Accessed 14 Feb 2019
38. Langmead B, Wilks C, Antonescu V et al (2019) Scaling read aligners to hundreds of threads on general-purpose processors. Bioinformatics 35:421–432
39. Li H, Handsaker B, Wysoker A et al (2009) The sequence alignment/map format and SAMtools. Bioinformatics 25:2078–2079
40. Hwang B, Lee JH, Bang D (2018) Single-cell RNA sequencing technologies and bioinformatics pipelines. Exp Mol Med 50:96
41. Bullard JH, Purdom E, Hansen KD et al (2010) Evaluation of statistical methods for normalization and differential expression in mRNA-Seq experiments. BMC Bioinformatics 11:94
42. Dillies M-A, Rau A, Aubert J et al (2013) A comprehensive evaluation of normalization methods for Illumina high-throughput RNA sequencing data analysis. Brief Bioinform 14:671–683
43. Evans C, Hardin J, Stoebel DM (2018) Selecting between-sample RNA-Seq normalization methods from the perspective of their assumptions. Brief Bioinform 19:776–792
44. StatQuest (2015) RPKM, FPKM and TPM, Clearly explained. https://www.rna-seqblog.com/rpkm-fpkm-and-tpm-clearly-explained/. Accessed 20 Feb 2019
45. Wickham H (2016) Elegant graphics for data analysis. In: Springer (ed) ggplot2, 2nd edn. Verlag, New York
46. Zhao S, Guo Y, Sheng Q et al (2014) Advanced heat map and clustering analysis using Heatmap3. Biomed Res Int 2014:986048. https://doi.org/10.1155/2014/986048
47. Frommer M, McDonald LE, Millar DS et al (1992) A genomic sequencing protocol that yields a positive display of 5-methylcytosine residues in individual DNA strands. Proc Natl Acad Sci 89:1827–1831
48. Gruntman M, Novoplansky A (2004) Physiologically mediated self/non-self discrimination in roots. Proc Natl Acad Sci U S A 101:3863–3867

Chapter 8

Genome-Wide Identification of Regulatory DNA Elements in Crop Plants

Zhanjie Li and Kai Wang

Abstract

cis-regulatory DNA elements (CREs) are noncoding but functional DNA sequences. The binding of regulatory proteins into CRE regions leads to chromatin high sensitive to DNase I digestion, which are termed as DNase I hypersensitive sites (DHSs). These DHSs can be efficiently detected through DNase I digestion followed by high-throughput DNA sequencing (DNase-seq). Thus, DNase-seq has become a powerful technique for DHSs mapping at whole-genome level in both plants and animals. Here we describe a DNase-seq procedure modified and developed for crop plants. These plants usually contain large amounts of repetitive sequences and complex organic constituents. With the main improvement in nuclei isolation, this method has been successfully used in mapping DHSs in cotton and sugarcane.

Key words DNase-seq, Regulatory DNA elements, Open chromatin, Transcription factor occupancy

1 Introduction

As the pace of genome sequencing has been significantly accelerated, more and more plant genomes have been sequenced in the last few years. However, the transcriptional control of biological processes, like development, differentiation, and cell response to the environment, is largely unknown. These processes are orchestrated and regulated by the binding of transcription factors to *cis*-regulatory DNA elements (CREs), mainly promoters and enhancers in plants [1]. DNase I hypersensitive site (DHS) mapping has historically recognized as a powerful tool for genome-wide identification of different types of regulatory elements [2]. DHSs identification is traditionally carried out using Southern blot hybridization-based approach [3, 4], and subsequently followed by microarray-based method (DNase-chip) [5, 6]. Recently, high throughput sequencing-based genome-wide DHS mapping (DNase-seq) has been developed in several model animal species, including yeast [6, 7], human [5, 8], and Drosophila melanogaster

Luis M. Vaschetto (ed.), *Cereal Genomics: Methods and Protocols*, Methods in Molecular Biology, vol. 2072,
https://doi.org/10.1007/978-1-4939-9865-4_8,

[9]. The DNase-seq method has obvious advantages over the Southern-blot or microarray-based approach, such as high-throughput and no need of prior information. Many valuable information has been obtained using this method for identifying of *cis*-regulatory DNA elements in model eukaryotes.

The assessment of chromatin accessibility in plants starts late, only limited successful applications genome-wide DHS mapping using DNase-seq method were reported in plants [10–14], especially in crop plants. This might attribute to the technical difficulty itself and technique challenge as for plant with large genome size and complex organic constituents of crop plants. In our lab, we have established the DNase-seq procedure to fit for crop plants with an improvement in nuclei isolation, and successfully make a genome-wide DHS mapping in cotton using this method.

2 Materials

2.1 Plants

Any crop plants with a sequenced genome can be used for DNase-seq analysis. Plant tissues at different biological processes like differentiation, development, and response to the environment can be used for DNase-seq experiments.

2.2 Nuclei Isolation

1. Corning conical tubes (50 mL).
2. Miracloth (Calbiochem).
3. Centrifuge with swing bucket rotor with cooling system.
4. Spatula.
5. Nylon paintbrush.
6. Nuclei isolation buffer (NIB): 10 mM Tris–HCl, 80 mM KCl, 10 mM EDTA, 1 mM spermidine, 1 mM spermine, 0.5 M sucrose, pH 9.5. Store at 4 °C. Supplement with 0.15% (v/v) β-mercaptoethanol and 1.0 mg/mL polyvinyl pyrrolidone (PVP) before use (*see* **Note 1**).
7. Nuclei washing buffer (NWB): Freshly add Triton X-100 into NIB to a final concentration of 0.5% (v/v) to prepared NWB (*see* **Note 2**).
8. Nuclei digestion buffer (NDB): 10 mM Tris–HCl, 10 mM NaCl, 3 mM $MgCl_2$, pH 7.4. Store at 4 °C.

2.3 DNase I Digestion, DNA Plug Preparation and Pulsed-Field Gel Electrophoresis (PFGE)

1. Eppendorf tubes (1.5 mL, 2.0 mL).
2. DNA plug modes (Bio-Rad).
3. Metal baths preset to 37 °C, 75 °C, and 45 °C.
4. Pulsed-field gel electrophoresis (PFGE) (CHEF-DRII system, Bio-Rad).
5. Gel staining shaker.

6. 10 U/μL DNase I (RNase-free) and 10× incubation buffer (Roche).
7. Stop solution: 50 mM EDTA, pH 8.0.
8. InCert low-melt agarose (Lonza).
9. Seakem® HGT agarose (Lonza).
10. LIDS buffer: 1% (w/v) LIDS (lithium dodecyl sulfate, Sigma), 10 mM Tris–HCl, pH 8.0, 100 mM EDTA.
11. 0.5× TBE buffer: 45 mM Tris-borate, 1 mM EDTA, pH 8.3.
12. Yeast chromosome PFG marker (NEB).

2.4 DNase-Seq Library Construction

1. Bench-top centrifuge for 1.5- and 2.0-mL microcentrifuge tubes.
2. Metal baths preset to 65 °C, 95 °C.
3. Thermocycler.
4. TBE PAGE gels (4–20%, premade) and vertical PAGE electrophoresis system (Bio-Rad).
5. Spin-X filter (Fisher).
6. Dynabeads® M-280 Streptavidin beads and Dynabeads® MPC®-S Magnetic Particle Concentrator (Invitrogen).
7. Razor blade.
8. Eppendorf tubes (0.5 mL, 1.5 mL, 2.0 mL).
9. Thermo Labquake™ rotator.
10. T4 DNA polymerase (NEB).
11. T4 DNA polymerase buffer: 50 mM NaCl, 10 mM Tris–HCl, pH 8.0, 10 mM $MgCl_2$, 1 mM dithiothreitol (DTT). Store at 4 °C, add DTT before use.
12. 10 mM dNTP (Roche).
13. 10× NEB Buffer 2 (NEB).
14. 10× NEB Buffer 4 (NEB).
15. DNA plug dissolving buffer: 10 mM Tris–HCl, pH 8.0, 100 mM NaCl.
16. 10 mM Tris–HCl buffer, pH 8.0.
17. Tris-saturated pure phenol solution, (pH 8.0) (Sigma).
18. Chloroform.
19. Glycogen.
20. 3 M NaOAC (pH 5.2).
21. 10 mg/mL ethidium bromide solution (Invitrogen).
22. Ice-cold ethanol.
23. 70% ethanol.

24. 5 U/μL T4 DNA ligase (Roche).
25. MmeI (NEB).
26. 500 μM S-adenosyl methionine (SAM) (NEB).
27. rApid alkaline phosphatase (Roche).
28. 1× TE buffer: 10 mM Tris–HCl pH 8.0, 0.1 mM EDTA.
29. 2× B&W buffer: 10 mM Tris–HCl, pH 7.5, 1.0 mM EDTA, 2 M NaCl.
30. 0.15 N NaOH (freshly prepared).
31. KAPA HiFi HotStart PCR Kit (Kapa Biosystems).
32. 10 bp DNA ladder (Invitrogen).
33. 6× Blue/orange loading buffer (Promega).
34. 21 gauge needle (BDMedical).
35. 2100 Bioanalyzer (Agilent Technologies).
36. Illumina Hiseq 2000/2500.
37. The oligos sequences for adaptor I and adaptor II (*see* **Note 3**):

    Adaptor I: (the underlined and bold sequence is the recognition site for restriction enzyme MmeI; and several modifications are used, *see* **Note 4**).

    Oligo1: 5′-Bio-ACAGGTTCAGAGTTCTACAG**<u>TCCGAC</u>** -3′.

    Oligo2: 5′-P-**<u>GTCGGA</u>**CTGTAGAACTCTGAAC-Amm-3′.

    Adaptor II: (the barcode sequences are showed in italic and bold; *see* **Note 5**).

    Oligo1:5′-P-***CTTGTA***TCGTATGCCGTCTTCTGCTTG-3′.

    Oligo2: 5′-CAAGCAGAAGACGGCATACGA***TACAAG***NN -3′ (N represents any of A, T, G, and C; *see* **Note 6**)
38. Primer sequences for adaptor I- and II-ligated DNA fragments amplification:

    Primer 1: 5′-CAAGCAGAAGACGGCATACGA-3′.

    Primer 2: 5′-AATGATACGGCGACCACCGACAGGTTCA GAGTTCTACAGTCCGA-3′.

3 Methods

3.1 Isolation of Nuclei

1. Grind the plant tissue into fine powder in liquid nitrogen and then transfer into a 50 mL ice-cold Corning centrifuge tube, each tube with about 10 mL powder. Store the powder at −80 °C if not used immediately (*see* **Note 7**).
2. Add the similar volume of ice-cold NIB to suspend the powder. Stirring the powder using a chilled spatula when adding the liquid.

3. Gently agitate the tube on ice for about 5–10 min to completely mix the powder, then filter through four-layer Miracloth into a new 50 mL Corning tube.
4. Centrifuge the filtered solution at 1000 × *g* × 10 min at 4 °C, and slowly pour off the supernatant as much as possible.
5. Wash the pellet 2–3 times with cold NWB (10 mL NWB for each washing). The pellet can be resuspend using soft nylon paintbrush and centrifuge at 1000 × *g* × 10 min for each time.
6. Add 10 mL of NDB to equilibrate the nuclei and centrifuge at 1000 × *g* × 10 min at 4 °C, and remove the supernatant completely.

3.2 DNase I Digestion and Identifying Optimal DNase Digestion by Pulsed-Field Gel Electrophoresis (PFGE)

1. Gently resuspend nuclei in 800 μL of cold NDB using paintbrush (*see* **Note 8**).
2. Slowly pipet 80 μL of the nuclei suspension into 6–10 prechilled 1.5 mL Eppendorf tubes by using wide bore tips. Gently swirl before each transfer (*see* **Note 9**).
3. Place tubes with nuclei in 37 °C metal bath and allow the nuclei to equilibrate for 1 min.
4. Add various amount of DNase I to each tube with the precise enzyme amounts specified (*see* Fig. 1) and mix well by gently inverting the tubes several times (*see* **Note 9**).
5. Incubate the tubes at 37 °C for 10 min, and gently invert during incubation. **Steps 4–5** should be timed carefully.
6. Stop the reactions by adding 80 μL of stop solution, mix well by inverting and place the tubes on ice. Now the total volume in each tube is approximately 170 μL.
7. Prepare 2 mL of 2% InCert low-melt agarose in stop solution, melt it at 75 °C in metal bath and switch to 45 °C after completely melted.
8. Equilibrate the tubes of the DNase I-digested nuclei at 45 °C for 1 min. Slowly pipet 170 μL/tube of melt gel agarose into each tube. Invert several times to mix completely. Slowly transfer 80 μL of the mixture into Bio-Rad plug molds using pipet with wide bore tips, and set it at 4 °C for solidification. The volume of each plug is about 80 μL, so each DNase I digestion sample will require four DNA plugs.
9. Carefully release the solidified DNA plugs into 50 mL Corning tube containing 45 mL of LIDS buffer (keep plugs with different DNase I concentrations in separate Corning tubes), and gently shake at 60 rpm for 1–2 h at room temperature (*see* **Note 10**).
10. Refresh the LIDS buffer and incubate tubes overnight at 37 °C without shaking.

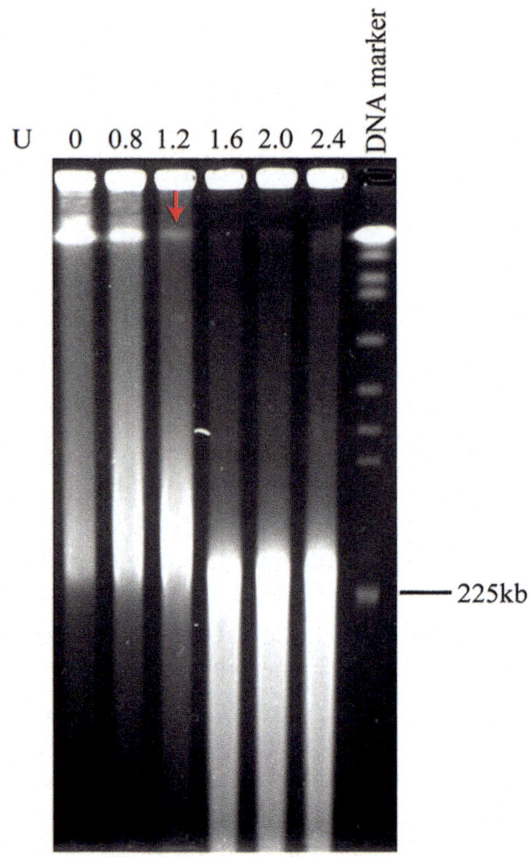

Fig. 1 Pulsed-field gel electrophoresis (PFGE) of cotton chromatin partially digested by DNase I. The nuclei were isolated from cotton leaves and digested by gradual amount of DNase enzyme for 10 min at 37 °C. The sample digested with 1.2 U DNase I (red arrow) is considered as the optimal digestion

11. Decant the LIDS buffer as much as possible, and wash plugs using 50 mL of stop solution at room temperature by gently shaking the tubes at 60 RPM for 1 h. Five times or more washing are recommended to ensure no visible detergent-caused bubbles are left after the final wash.
12. Store the DNA plugs in stop solution at 4 °C for later use or run PFGE immediately as follows.
13. Prepare 170 mL of 1% agarose using 0.5× TBE and pour into the PFGE-specific gel mold (Bio-Rad). Load one half of one plug from each DNase I concentration into each well. Using NEB yeast chromosome PFG marker to size smearing pattern.
14. Run PFGE (CHEF-DRII system) with the following parameters: 20–60 s switch time for 17.5 h, 6 V/cm (180 V total) (*see* Fig. 1). Keep 0.5× TBE running buffer chilled at 10 °C. After

electrophoresis, the gel was stained in water with 0.5 μg/mL EtBr for 0.5–1 h to visualize the smearing profiles under UV light.

3.3 DNase-Seq Library Construction

3.3.1 Blunt Ending DNase-Digested Ends

1. Choose DNA plugs from optimal concentrations for blunt ending treatment (*see* **Note 11**).
2. Wash the four plugs from each chosen concentration within 50 mL of 1× T4 DNA polymerase buffer to completely remove the EDTA (two times for 1 h each time), shaking at 60 rpm at room temperature.
3. Discard the solution from the 50 mL Corning tube and keep the DNA plugs at the bottom of the tubes.
4. Incubate the DNA plugs with T4 DNA polymerase/buffer in 50 mL Corning tube at room temperature for 4 h to completely trim the DNA ends (*see* **Note 12**). Gently invert the tubes occasionally. The reaction systems are as follows:

   DNA plug: 80 μL.

   10× NEB Buffer 2: 12 μL.

   10 mM dNTP: 5 μL.

   T4 DNA polymerase (NEB): 6 μL.

   100× BSA: 2 μL.

   ddH_2O: 99.2 μL.

   Total volume: 204.2 μL

   Each component should be scaled up accordingly if multiple DNA plugs need to be trimmed.
5. After reaction, rinse the plugs quickly with 10 mL of 1 × T4 DNA polymerase buffer to remove residual enzyme.
6. Transfer each plug to 500 μL of 10 mM Tris–HCl, pH 8.0,100 mM NaCl and incubate at 65 °C for 15 min to melt the plugs. Flip each tube every 5 min to make sure the agarose is dissolved completely.
7. Sequentially extract the DNA with an equal volume of phenol, phenol-chloroform (1:1), and chloroform, respectively. Precipitate DNA with two volumes of ice-cold ethanol, 1/10 volume of 3 M NaOAC (pH 5.2), and 1 μL glycogen by incubating at −20 °C for 10 min and centrifuging at 18,506 × g for 16 min at 4 °C.
8. Wash DNA pellet with 500 μL of 70% ethanol and centrifuge at 18,506 × *g* for 5 min, remove all residual liquid and air-dry the DNA for 3–4 min at room temperature.
9. Resuspend DNA pellet in 15 μL of 10 mM Tris–HCl (pH 8.0) per DNA plug. This DNA solution can be stored at −20 °C for future use or immediately go to the following steps.

3.3.2 Ligation of Adaptors

1. Anneal the oligos to make adaptor I and adaptor II before ligate the blunt-DNA to adaptors (*see* **Note 3**).
2. Ligation blunt-DNA to adaptor I using T4 DNA ligase by incubating the tubes at 20 °C overnight. The reaction systems are as follows:

   Blunt-end DNA: 12 μL.

   10× ligation buffer: 5 μL.

   5 U/μL T4 DNA ligase (Roche): 2 μL

   25 pmol/μL adaptor I: 6 μL.

   ddH_2O: 25 μL.

   Total volume: 50 μL
3. Recovery of ligated DNA with adaptor I using gel purification. Prepare 0.8% low melt gel using 1× TBE containing 10 mM EDTA (*see* **Note 13**). Run the DNA samples from above reactions for 40 min at 80 V in 1× TBE running buffer supplemented with 10 mM EDTA (*see* Fig. 2).
4. Cut out the high molecular weight (HMW) DNA bands at the top of the gel. Transfer gel pieces from the same sample into 500 μL of DNA plug dissolving buffer, and melt it completely at 65 °C for 15 min. Gently flip every few minutes.

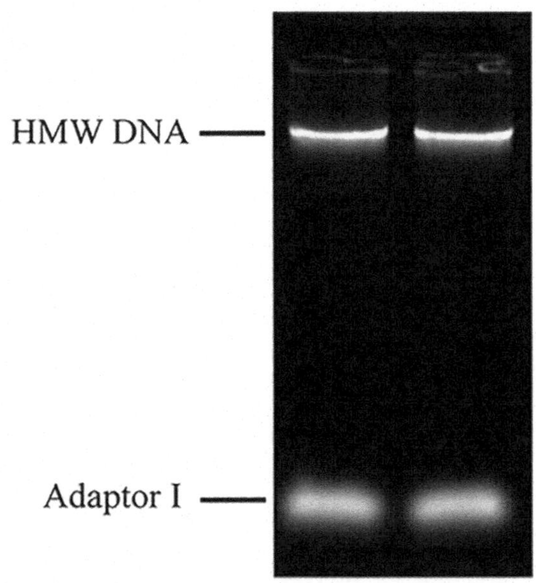

Fig. 2 Low melting agarose gel electrophoresis of high molecular weight (HMW) DNA ligated with adaptor I. The HMW-DNA can be separated from free adaptor I and appear as a sharp single band. This sharp band should be cut out for DNase-seq library preparation

5. Extract the ligated HMW DNA with the same procedures as **steps 7–9** in Subheading 3.3.1. Finally, resuspend the DNA pellet in 75 μL of 10 mM Tris–HCl (pH 8.0). The DNA solution can be stored at −20 °C or immediately used for the next steps.
6. Trim the adaptor I-ligated DNA fragments at 37 °C for 1.5 h by using MmeI (*see* **Note 14**). Prepare the MmeI digestion system as follows:

   Adapter I-ligated HMW DNA: 75 μL.

   10× NEB Buffer 4: 10 μL.

   500 μM SAM (S-adenosyl methionine): 10 μL.

   MmeI: 5 μL.

   Total volume: 100 μL
7. After MmeI digestion, add 3 μL of rApid alkaline phosphatase and continue to incubate at 37 °C for additional 1 h.
8. Extract the digestion with the same procedures as **steps 7–9** in Subheading 3.3.1, and finally resuspend the DNA in 50 μL of ultrapure water.
9. Pipet 100 μL of M-280 streptavidin beads for each sample into a 1.5 mL Eppendorf tube (mix well before pipet), put the tubes on magnetic bead concentrator and wait for 1 min to separate the beads from storage solution.
10. Carefully remove and discard the supernatant and wash the beads two times with 1 mL of 1× TE and one time with 1 mL of 1× B&W buffer. Each time using the magnetic bead concentrator to collect the beads and discard the wash solution.
11. Add 50 μL of DNA obtained in **step 8** and 50 μL of 2× B&W buffer to resuspend the beads, incubate at 30 °C for 30 min, gently flip every 10 min.
12. Place the tubes on the magnetic bead concentrator to separate the beads from the supernatant, and remove the supernatant as much as possible.
13. Wash the beads three times with 1 mL of 1× TE, and remove the supernatant as much as possible after the last time.
14. Ligation of barcoded adaptor II to adaptor I-ligated DNA fragments on M-280 beads by rotating at room temperature for 4 h. Prepare the ligation reaction system as follows:

    DNA-bound M-280 beads: approximately 10 μL.

    10× T4 DNA ligase buffer: 10 μL.

    25 pmol/μL adaptor II (prepared in **step 1**): 6 μL.

    5 U/μL T4 DNA ligase (Roche): 2 μL.

    ddH_2O: 72 μL.

    Total volume: 100 μL

3.3.3 PCR Enrichment of Adaptors-Ligated DNA Fragments

1. After adaptor II ligation, wash the DNA-bound beads with 1 mL of 1× TE and remove the supernatant as much as possible.
2. Add 500 μL of freshly prepared 0.15 N NaOH to the beads and rotate the tubes at room temperature for 5 min, then completely remove the NaOH solution and wash the beads five times with 1 mL of 1× TE (*see* **Note 15**).
3. Remove the TE as much as possible after the last wash, and resuspend the beads in 20 μL of 10 mM Tris–HCl (pH 8.0).
4. Amplify the adaptor I- and II-ligated DNA fragments using PCR reactions (*see* **Note 16**).

   Prepare the following PCR reaction system:

   DNA-bound beads suspension: 10 μL.

   25 μM PCR primer 1: 0.5 μL.

   25 μM PCR primer 2: 0.5 μL.

   KAPA HiFi Hotstart PCR ReadyMix: 20 μL.

   ddH_2O: 9 μL.

   Total volume: 40 μL

   Run the following PCR program: 98 °C for 30 s; 12 cycles of 98 °C for 10 s, 60 °C for 30 s, and 72 °C for 15 s; final extend 72 °C for 7 min.

3.3.4 Purification of PCR Products Using Polyacrylamide Gel Electrophoresis (PAGE)

1. Premix 40 μL PCR products with 8 μL loading buffer. Load 1.2 μL of 10 bp DNA marker into one well, and split 48 μL of PCR products into 2 adjacent wells of 4–20% precast TBE PAGE gel (Bio-Rad).
2. Run the PAGE gel at 120 V for about 2 h in the 1× TBE running buffer, after electrophoresis, pry apart the gel cassette and stain the gel in 50 mL of 1× TE plus 2.3 μL of 10 mg/mL ethidium bromide for 2–3 min. Observe the gel under the UV light (*see* Fig. 3).
3. Cut out the DNA band corresponding to ~90 bp (adaptors + insert) using a clean razor blade, and place the gel piece into a 0.5-mL Eppendorf tubes whose bottom has a couple of tiny holes punched by a 21 gauge needle.
4. Set the 0.5 mL tube into a 2-mL Eppendorf collection tube with round bottom, and spin at 18,506 × *g* for 2.5 min to crush the gel through the hole.
5. Add 150 μL of 1× NEB Buffer 2 to the gel and elute the DNA by rotating the tube gently at room temperature for 2 h using Thermo Labquake™ rotator.

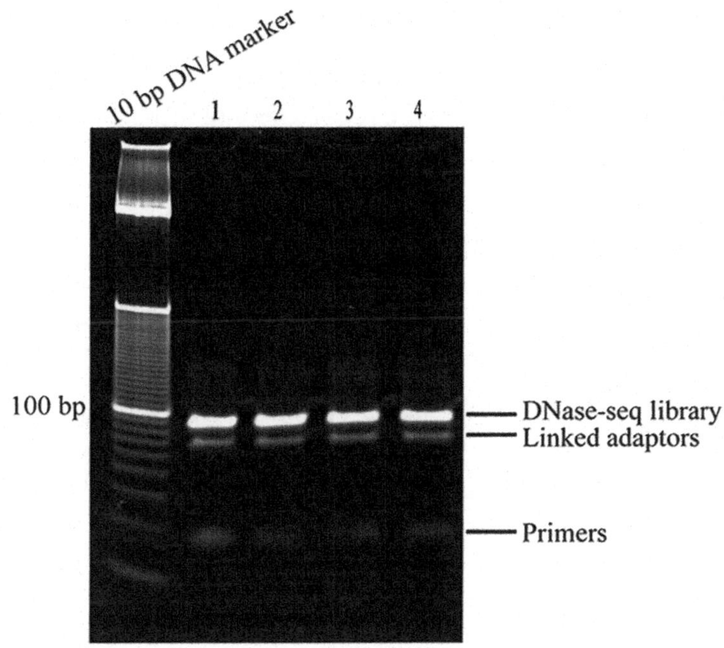

Fig. 3 Polyacrylamide gel electrophoresis (PAGE) of the final PCR reaction. Approximately 1 μL of 10 bp ladder and 25–30 μL of the PCR product was loaded in each well of the 4–20% PAGE gel. The DNA band ~90 bp is interest, which contains the genomic DNA insert (~20 bp) and adaptors (totally 66 bp), and the other two smaller bands were linked adaptors (66 bp) and PCR primers (20–30 bp)

6. Pipet the eluate and the gel debris mixture onto the top of a Spin-X filter using a wide bore tip, spin at 1200 × *g* for 5 min for completely remove gel and collect the filtered solution into a new 1.5 mL Eppendorf tube.
7. Extract the DNA with the same procedures as **steps 7–9** in Subheading 3.3.1, and finally resuspend the DNA in 13 μL of 10 mM Tris–HCl (pH 8.0).
8. Rerun 1 μL of recovered DNA on 4–20% TBE PAGE gel to verify the quality of the DNase-seq library.

3.3.5 DNase-Seq and DHSs Mapping

1. Check the libraries quality using the BioAnalyzer before sequencing. And then perform high-throughput sequencing of DNase-seq libraries on the Illumina GAII or HiSeq 2000/2500 platform. Using the custom sequencing primer (*see* **Note 17**) (5′-CCACCGACAGGTTCAGAGTTCTA CAGTCCGAC-3′) to perform single-end sequencing of libraries, according to the manufacture's recommendation.

2. Split the datasets using barcode identifiers if multiple libraries are sequenced in the same lane. Remove adaptors using cutadapt [15] or other programs, and then align the reads to the reference genome using Bowtie [16] or BWA [17] with 1 bp or no mismatch allowed. Identify DHSs (DNase I hypersensitive sites) using F-seq [2] with appropriate parameters (recommended with 200–300 bandwidth and FDR less than 0.05 or 0.01).

4 Notes

1. Some crop plant tissues usually contain large amounts of polysaccharide and polyphenol, which can severely influence nuclei isolation and DNase I digestion. The addition of polyvinylpyrrolidone (PVP) can remove polysaccharide and polyphenol efficiently during nuclei isolation.
2. Triton X-100 is added to remove organelle contamination (including chloroplast and mitochondria) by breaking the membrane. However, it is recommended that the maximum concentration of Triton X-100 should not be more than 0.5% (v/v), since high concentration of Triton X-100 can cause degradation of nuclei/chromatin and increase DNase-seq background noise.
3. The oligos for adapter I and II are stocked at the concentration of 25 pmol/μL. Before uses, two oligos for each adaptor are annealed in 1× NEB Buffer 2, by heating at 95 °C for 5 min and slowly cooling to room temperature. The aliquots of the annealed adaptors can be kept at −20 °C.
4. Biotin, Amino (Amm), and phosphorylation (P) modifications are incorporated in the oligos of adaptor I as indicated. Biotin is used for streptavidin-based purification of adaptor-ligated DNA fragments. Amino modifications are typically used for attachment of other modifiers (such as fluorescent dyes) or for attaching an oligo to a solid surface. And 5′ phosphorylation is used as a substrate for T4 DNA ligase to ligate adaptor I with blunt-end DNase I-trimmed DNA fragments.
5. For adaptor II, we provide an example of barcoded sequence (six codes are in italic and bold) from Illumina Truseq index. In fact, the total 12 Truseq adapters released from Illumina (ATCACG, CGATGT, TTAGGC, TGACCA, ACAGTG, GCCAAT, CAGATC, ACTTGA, GATCAG, TAGCTT, GGCTAC, and CTTGTA) can all be used as barcodes.
6. The 5′ overhanging NN in adaptor II will serve to facilitate the ligation with MmeI-digested DNA molecules with a 3′ overhanging dinucleotide.

7. The yield of nuclei highly dependents on the quality (young or old) of tissues. Fresh young tissues are always preferred and recommended. In general, the recommended 10 mL of plant powder will be sufficient to generate enough nuclei for most tissue types. Plant tissues should be ground into fine powder in liquid nitrogen to thoroughly disrupt cell walls and other fibrous structures and to ensure maximum yield of nuclei.
8. The volume of NDB for resuspending the nuclei depends on the amount of the tissue powder and the yield of nuclei, it should be scaled up or down according to the variation in different experiments.
9. A key step in the DNase-seq assay is to determine the concentration of enzyme that results in the releasing of the most accessible compartments of the genome (DNase I hypersensitive sites) while minimizing digestion of bulk chromatin. DNase I concentrations that are too low or too high can lead to sporadic digestion or over digestion of chromatin. Therefore, at least 4–5 DNase I concentrations are recommended to find the optimization of digestion. Furthermore, the ideal DNase I concentrations can vary significantly between different tissue types or species. A recommended strategy is to set vary enzyme concentrations with a fixed digestion time.
10. LIDS is used to denature proteins within the digested chromatin to release high molecular weight (HMW) DNA. This step should be carried out in the plugs to prevent artificial shearing of the HMW DNA.
11. Since DHSs are not binary, but instead represent a continuum of signal intensities, multiple optimal concentrations for DNase-seq can be included to capture both the strong and weak DHSs. As shown in Fig. 1, the optimal concentrations exhibit a smearing DNA pattern with a small percentage of undigested large high molecular weight (LHMW) DNA present at the top of the gel (1.2 U). In contrast, over-digested or under-digested samples contain no visible or a larger portion of undigested LHMW DNA.
12. Since the DNase I nicks DNA and leaves single-strand overhangs, these overhangs need to be blunt ended by T4 DNA polymerase before ligating to blunt-ended adaptor I. This process is accomplished by using 5′–3′ polymerase activity to fill the 5′ overhangs and 3′–5′ exonuclease activity to trim the 3′ overhangs, respectively.
13. 10 mM EDTA should be added into both the gel and running buffer to prevent possible degradation of HMW DNA during electrophoresis.

14. MmeI is a type II restriction enzyme that cuts 20 nt from the recognition sequence and generates small DNA fragments containing DNase I cutting sites suitable for Illumina sequencing library preparation.
15. Alkaline treatment is used to denature double-stranded DNA template on the beads and remove the nonbiotinylated strand. Since the 2 bp overhang generated by MmeI does not always perfectly match up with the end of the adaptor II. Thus, getting rid of nonbiotinylated strand will help to eliminate false-positive sequencing of these two mismatched base pairs.
16. Any MmeI-treated short DNA fragments flanked with both adapter I and adapter II can be amplified by PCR to reach about 50–100 nM in the final purified solution.
17. Please refer to Illumina recommendations for additional details regarding sequencing.

Acknowledgments

This work was supported by the National Natural Science Foundation of China (31771862), Scientific Research Foundation of the Education Department of Fujian Province (JK2017016), and Natural Science Foundation of Fujian Province (2018J01711).

References

1. Jiang J (2015) The 'dark matter' in the plant genomes: non-coding and unannotated DNA sequences associated with open chromatin. Curr Opin Plant Biol 24:17–23
2. Boyle AP, Davis S, Shulha HP et al (2008) High-resolution mapping and characterization of open chromatin across the genome. Cell 132(2):311–322
3. Wu C (1980) The 5′ ends of *Drosophila* heat shock genes in chromatin are hypersensitive to DNase I. Nature 286(5776):854–860
4. Kodama Y, Nagaya S, Shinmyo A et al (2007) Mapping and characterization of DNase I hypersensitive sites in *Arabidopsis* chromatin. Plant Cell Physiol 48:459–470
5. Sabo PJ, Humbert R, Hawrylycz M et al (2004) Genome-wide identification of DNaseI hypersensitive sites using active chromatin sequence libraries. Proc Natl Acad Sci U S A 101:4537–4542
6. Sabo PJ, Kuehn MS, Thurman R et al (2006) Genome-scale mapping of DNase I sensitivity in vivo using tiling DNA microarrays. Nat Methods 3:511–518
7. Hesselberth JR, Chen X, Zhang Z et al (2009) Global mapping of protein-DNA interactions *in vivo* by digital genomic footprinting. Nat Methods 6:283–289
8. Crawford GE, Holt IE, Mullikin JC et al (2004) Identifying gene regulatory elements by genome-wide recovery of DNase hypersensitive sites. Proc Natl Acad Sci U S A 101:992–997
9. Thomas S, Li XY, Sabo PJ et al (2011) Dynamic reprogramming of chromatin accessibility during *Drosophila* embryo development. Genome Biol 12:R43
10. Zhang W, Zhang T, Wu Y et al (2012) Genome-wide identification of regulatory DNA elements and protein-binding footprints using signatures of open chromatin in *Arabidopsis*. Plant Cell 24:2719–2731
11. Zhang W, Wu Y, Schnable JC et al (2012) High-resolution mapping of open chromatin in the rice genome. Genome Res 22:151–162
12. Sullivan AM, Arsovski AA, Lempe J et al (2014) Mapping and dynamics of regulatory DNA and transcription factor networks in *A. thaliana*. Cell Rep 8:2015–2030

13. Qiu Z, Li R, Zhang S et al (2016) Identification of regulatory DNA elements using genome-wide mapping of DNase I hypersensitive sites during tomato fruit development. Mol Plant 9:1168–1182
14. Zhao H, Zhang W, Chen L et al (2018) Proliferation of regulatory DNA elements derived from transposable elements in the maize genome. Plant Physiol 176:2789–2803
15. Martin M (2011) Cut adapt removes adapter sequences from high-throughput sequencing reads. EMBnet J 171(17):10–12
16. Langmead B, Trapnell C, Pop M et al (2009) Ultrafast and memory-efficient alignment of short DNA sequences to the human genome. Genome Biol 10:R25
17. Li H, Durbin R (2009) Fast and accurate short read alignment with burrows-wheeler transform. Bioinformatics 25:1754–1760

Chapter 9

Genome-Wide Profiling of Histone Modifications with ChIP-Seq

William A. Ricci, Laura Levin, and Xiaoyu Zhang

Abstract

Chromatin immunoprecipitation coupled with sequencing (ChIP-seq) is a widely used method for mapping the genome-wide locations of chromatin-associated proteins. This protocol has been developed and utilized to perform ChIP on histone covalent modifications in various plant species including cereals. DNA and chromatin-associated proteins are crosslinked with formaldehyde. Chromatin is then isolated from nuclei and sheared via sonication. Antibodies targeting the histone modification of interest are incubated with the sheared chromatin and nonspecific interactions are washed away. DNA is purified via phenol-chloroform extraction, end-repaired, ligated to sequencing adapters, and PCR-amplified.

Key words Plant, Cereals, Chromatin immunoprecipitation, Chip-seq, Nuclei, Histone modification, Epigenomics, Antibody

1 Introduction

In eukaryotes, nuclear DNA wraps around octamers of core histone proteins to form nucleosomes, which act as the basic structural units of chromatin. Histones are subject to a wide array of post-translational covalent modifications that can influence local or higher-order chromatin structure or serve as platforms to recruit chromatin-binding proteins. Consequently, histone modifications are critical for many processes that occur within the chromatin environment, such as transcription, RNA processing, chromatin condensation, DNA replication, and DNA lesion repair [1–4].

Chromatin immunoprecipitation coupled with high-throughput sequencing (ChIP-seq) is the principle method for studying the genome-wide histone modification landscape [5–8]. In the described procedure, DNA-histone interactions are first fixed by formaldehyde-mediated crosslinking of plant tissue. Nuclei are extracted from the crosslinked tissue via a general extraction procedure which works on nuclei of various sizes and densities. Purified nuclei are lysed and the chromatin is released. Sonication

Luis M. Vaschetto (ed.), *Cereal Genomics: Methods and Protocols*, Methods in Molecular Biology, vol. 2072, https://doi.org/10.1007/978-1-4939-9865-4_9,

is used to fragment and solubilize the chromatin. The chromatin fragments are incubated with antibodies targeting the epitopes of interest and chromatin fragments lacking the desired epitopes are washed away. Following immunoprecipitation, the chromatin is reverse-crosslinked and the DNA is purified via a standard phenol-chloroform purification. The sheared DNA ends are repaired, sequencing adapters ligated to ends, and DNA fragments PCR-amplified. Following sequencing and bioinformatic analyses, the locations, relative abundances, and quantitative changes of histone modifications can be determined in a genome-wide manner.

This chapter describes the experimental procedures of a general ChIP-seq method that has been successfully implemented in a variety of plant species with genome sizes ranging from ~156 Mb (*Arabidopsis thaliana*) to ~5 Gb (*Hordeum vulgare*) and on a variety of tissues, such as leaves, roots, endosperm, and immature inflorescences.

2 Materials

Following are the required materials and buffers for different parts of the procedure and the quantities required for a single experiment. Some buffers should be prepared as stocks and the unstable components added immediately before use (*see* **Note 1**).

2.1 Crosslinking Tissue

1. Vacuum chamber capable of reaching −25″ Hg.
2. Forceps.
3. Razor blade.
4. Square of Miracloth.
5. Twist tie.
6. 50 mL falcon tube (or equivalent).
7. DI water.
8. 37% Formaldehyde.
9. 0.1 M PMSF stored in isopropanol at −20 °C.
10. Crosslinking buffer (15.4 mL): 1 mM EDTA, pH 8.0, 0.4 M sucrose, 10 mM Tris–HCl, pH 8.0. Store at 4 °C.
11. 1 M glycine.
12. Approximately 1 g of plant tissue (*see* **Note 2**).

2.2 Preparing Beads with Antibodies

1. Labquake tube rotator (or equivalent).
2. Magnetic 1.5 mL tube rack.
3. 1.5 mL tube.

Table 1
Commonly used antibodies for targeting histone modifications in plants

Epitope	Source	Cat #	μg antibody per 25 μL beads	Capture with protein A or G beads
H3	Abcam	ab1791	1.5	A
H3K4me1	Abcam	ab8895	1.5	A
H3K4me3	Millipore	07-473	1.5	A
H3K9ac	Active Motif	61251	1.5	G
H3K23ac	Millipore	07-355	1.5	A
H3K27me3	Millipore	07-449	1.5	A
H3K36me3	Abcam	ab9050	1.5	A
H3K56ac	Millipore	07-677-1	1.5	A

Selected histone modifications and antibodies commonly used to target them. All antibodies shown can be captured with both protein A and G beads; however the bead type listed shows stronger affinity and therefore is preferable

4. Dynabeads® Protein A (Thermo Fisher Scientific) or Dynabeads® Protein G (Thermo Fisher Scientific). *See* Table 1 for which beads to use with which antibodies.
5. Antibodies (*see* Table 1 and **Note 3**).
6. ChIP dilution buffer without protease inhibitor cocktail (3 mL): 1.2 mM EDTA, 167 mM NaCl, 16.7 mM Tris–HCl, pH 8.0, 1.1% (v/v) Triton X-100, 0.1 mM PMSF. Store at 4 °C.
7. ChIP dilution buffer with protease inhibitor cocktail (100 μL): 1.2 mM EDTA, 167 mM NaCl, 16.7 mM Tris–HCl, pH 8.0, 1.1% (v/v) Triton X-100, 0.1 mM PMSF, 1× cOmplete™ Mini EDTA-free Protease Inhibitor Cocktail. Store at 4 °C.

2.3 Nuclei Extraction and Chromatin Isolation

1. 4 °C bench-top centrifuge suitable for 1.5 mL tubes.
2. 50 mL falcon tubes.
3. Squares of Miracloth.
4. 1.5 mL tubes
5. Mortar and pestle.
6. Liquid N_2.
7. Extraction buffer 1(30 mL): 1 mM EDTA, 10 mM $MgCl_2$, 0.4 M sucrose, 10 mM Tris–HCl, pH 8.0, 5 mM BME, 0.1 mM PMSF, 1× cOmplete™ Mini EDTA-free Protease Inhibitor Cocktail. Store at 4 °C.

8. Extraction buffer 2 (1 mL): 1 mM EDTA, 10 mM $MgCl_2$, 0.25 M sucrose, 10 mM Tris–HCl, pH 8.0, 5 mM BME, 0.1 mM PMSF, 1× cOmplete™ Mini EDTA-free Protease Inhibitor Cocktail, 1% (v/v) Triton X-100. Store at 4 °C.
9. Extraction buffer 3 (600 μL): 1 mM EDTA, 2 mM $MgCl_2$, 1.7 M sucrose, 10 mM Tris–HCl, pH 8.0, 0.15% (v/v) Triton X-100, 0.1 mM PMSF, 1× cOmplete™ Mini EDTA-free Protease Inhibitor Cocktail. Store at 4 °C.
10. Nuclei lysis buffer (300 μL): 10 mM EDTA, 1% (v/v) SDS, 50 mM Tris–HCl, pH 8.0, 0.1 mM PMSF, 1× cOmplete™ Mini EDTA-free Protease Inhibitor Cocktail. Store at 4 °C.

2.4 Chromatin Sonication

1. 4 °C bench-top centrifuge suitable for 1.5 mL tubes.
2. Bioruptor® UCD-200 with chilling pump or ice.
3. 1.5 mL tubes.
4. 5 M NaCl.
5. Elution buffer (780 μL): 1% (v/v) SDS, 0.1 M $NaHCO_3$. Prepare fresh before use since $NaHCO_3$ is unstable in aqueous solution.

2.5 Chromatin Immunoprecipitation

1. 5 or 15 mL falcon tube.
2. Chip dilution buffer with protease inhibitor cocktail (2.34 mL).
3. Magnetic tube rack.
4. 1.5 mL tube.
5. Chip dilution buffer without protease inhibitor cocktail (3 mL).
6. Low salt wash buffer (2 mL): 2 mM EDTA, 150 mM NaCl, 0.1% (v/v) SDS, 20 mM Tris–HCl, pH 8.0, 1% (v/v) Triton X-100.
7. High salt wash buffer (2 mL): 2 mM EDTA, 500 mM NaCl, 0.1% (v/v) SDS, 20 mM Tris–HCl, pH 8.0, 1% (v/v) Triton X-100.
8. LiCl wash buffer (2 mL): 1 mM EDTA, 0.25 M LiCl, 1% (w/v) sodium deoxycholate, 1% (v/v) NP-40, 10 mM Tris–HCl, pH 8.0.
9. TE buffer (1 mL): 10 mM Tris–HCl, pH 8.0, 1 mM EDTA.
10. Elution buffer (400 μL).

2.6 Reverse Crosslinking

1. 5 M NaCl (for both sample and input).
2. Chip elution buffer (380 μL).

2.7 Proteinase Digestion

1. 0.5 M EDTA.
2. 1 M Tris–HCl, pH 6.5.
3. 20 mC/mL Proteinase K Solution (Applied Biosystems).

2.8 DNA Purification

1. 4 °C bench-top centrifuge suitable for 1.5 mL tubes.
2. 1.5 mL tube.
3. Phenol (buffered to pH 8.0).
4. Chloroform.
5. 2.5 M potassium acetate.
6. 100% EtOH.
7. 20 mg/mL glycogen.
8. 10 mM Tris–HCl, pH 8.0.
9. 70% EtOH.

2.9 DNA Sequencing Library Preparation (End-Repair, A-Tailing, Ligation, and PCR)

1. Magnetic rack suitable for 0.2 mL PCR tube.
2. Thermal cycler.
3. DNA end-repair master mix (22.85 μL): 7.14 μL 10× End-It buffer, 7.14 μL 10 mM dNTPs, 7.14 μL 10 mM ATP, and 1.43 μL End-It enzyme mix. End-It buffer and enzyme mix come from the End-It™ DNA End-Repair Kit (Lucigen). Prepare on ice immediately before use.
4. Ampure XP beads (Agencourt).
5. 80% EtOH.
6. 10 mM Tris–HCl, pH 8.0.
7. NEBNext® dA-Tailing Reaction Buffer (NEB).
8. Klenow fragment (3′ ≥ 5′ exo-, 5 U/μL) (NEB).
9. T4 DNA Ligase Reaction Buffer (NEB).
10. 400 U/μL T4 DNA Ligase(NEB).
11. 10 μM Illumina Truseq adapters.
12. PCR master mix (30 μL): 14.66 μL nuclease-free water, 3.34 μL 10 μM Illumina TruSeq primer mix, 10 μL 5× Phusion HF buffer (NEB), 1 μL 10 mM dNTPs, 1 μL 2 U/μL Phusion® High-Fidelity DNA Polymerase (NEB). Mix well by pipetting. Prepare on ice immediately before running PCR.

3 Methods

The entire procedure can be comfortably performed within 4 days. *See* **Note 4** for procedural time considerations.

3.1 Formaldehyde-Mediated Crosslinking of Tissues

All steps performed at room temperature.

1. Harvest approximately 1 g of plant tissue. This amount of starting material is usually sufficient for at least 6 ChIPs. However, depending on the tissue type, this amount can be reduced (*see* **Note 2**).

2. In a 50 mL Falcon (or equivalent) tube, combine 15.4 mL of crosslinking buffer with 160 μL of 0.1 M PMSF and 432 μL of 37% formaldehyde (*see* **Note 5**).
3. Dice the plant tissue into ~1 mm cross sections. Wrap up the chopped tissue in a single layer of Miracloth and close with a twist tie to form a "dim sum" bag (Fig. 1a).
4. Insert the dim sum bag into the crosslinking solution in the 50 mL tube. Use the twist tie to hold the dim sum bag below the surface of the solution (Fig. 1b).
5. Place the tube under a −25″ Hg vacuum and incubate for 10 min (*see* **Note 6**).
6. Release the vacuum and incubate for 5 min.
7. Swirl the tube approximately 10 times.
8. Reapply the −25″ Hg vacuum for 5 min.
9. Release the vacuum and add 1.6 mL of 1 M glycine to the tube. Invert the tube 10 times to mix. The glycine quenches the remaining formaldehyde.
10. Reapply the −25″ Hg vacuum for 5 min.

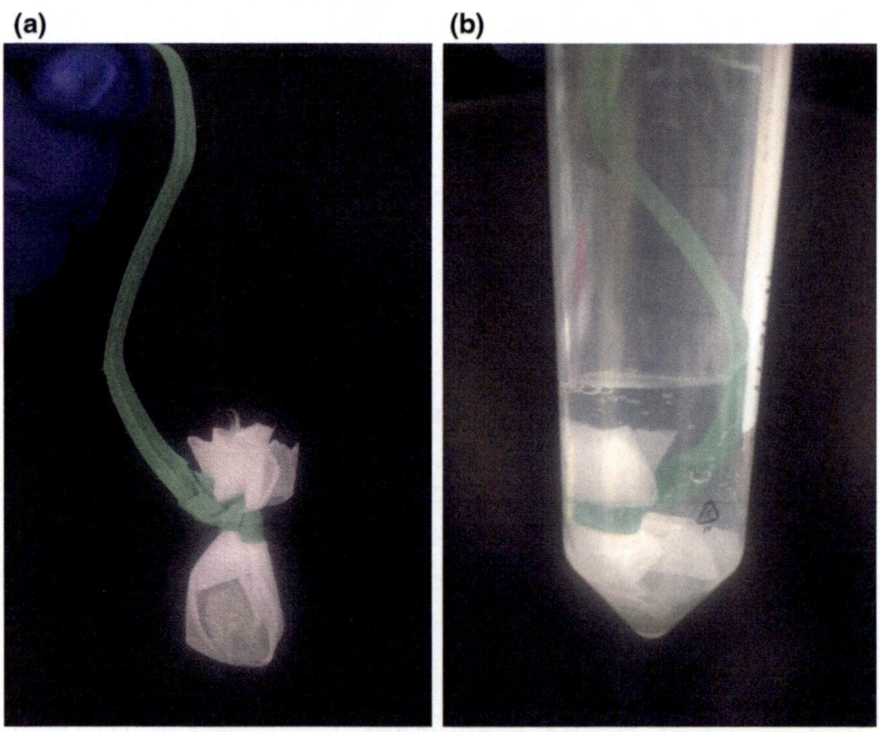

Fig. 1 The method for crosslinking tissue. (**a**) The "dim sum" bag, comprising Miracloth, a tie strap, and plant tissue inside. The tie strap acts as a convenient handle. (**b**) The dim sum bag submerged in crosslinking buffer in a 50 mL tube. The tie strap is angled to press against the tube and hold the bag below the surface

11. Discard the crosslinking solution in an appropriate hazardous waste container and wash the dim sum bag with ~50 mL of DI H_2O 5 times, each time moving the dim sum bag up and down 10 times.
12. Remove excess H_2O by blotting with a paper towel. Flash freeze crosslinked tissue in liquid nitrogen in Al foil and store at −80 °C. Excess water may make it more difficult to grind the tissues in future steps. Crosslinked tissue can be stored at −80 °C for several months before being used for ChIP.

3.2 Preparing Antibody-Coated Magnetic Beads

All steps should be performed at 4 °C.

1. Resuspend protein A or G Dynabeads by vortexing (*see* Table 1 for guidance on which beads to use with which antibodies). Place 25 μL (750 μg) of Dynabeads in a 1.5 mL Eppendorf tube.
2. Wash the beads 3 times with 1 mL of ChIP dilution buffer (containing water instead of protease inhibitor cocktail) each time. To wash beads, place tubes with beads on a magnetic rack and remove the Dynabead storage solution. Remove tubes from the magnetic rack, resuspend the beads with 1 mL of ChIP dilution buffer, and place the tubes back on the magnetic rack. Repeat the wash 2 more times.
3. Remove the supernatant and add 100 μL of ChIP dilution buffer (this time containing protease inhibitor cocktail) and 1.5 μg of antibody to the washed beads.
4. Incubate with rotation at 4 °C for 3 h to overnight.

3.3 Nuclei Extraction and Chromatin Isolation

All steps should be performed at 4 °C.

1. Grind crosslinked tissue in liquid nitrogen to a very fine powder. It generally takes ~4 min of grinding to reach a sufficient fineness.
2. Add powder to 30 mL of extraction buffer 1 in a 50 mL tube. Resuspend the powder by inverting the tube ~20 times and vortexing ~10 s.
3. Filter the suspension through two layers of miracloth into a new 50 mL tube. Gently squeeze the miracloth to push through the last of the filtrate.
4. Centrifuge the filtrate at 2,880 RCF for 20 min at 4 °C. This step pellets the nuclei and other organelles without bursting the nuclei.
5. Aspirate the supernatant and resuspend the pellet in 1 mL of extraction buffer 2. The pellet should resuspend with no more than 10 gentle pipettes up and down. The detergent concen-

tration within this extraction buffer selectively eliminates plastids and mitochondria.

6. Transfer the suspension to a 1.5 mL tube and centrifuge for 10 min at 12,000 RCF at 4 °C.
7. Remove the supernatant and resuspend the pellet in 300 μL of extraction buffer 3. This resuspension step requires more vigorous pipetting. Some particles may not resuspend, which is not problematic.
8. In a new 1.5 mL Eppendorf tube, add 300 μL of extraction buffer 3. Slowly add the 300 μL of resuspended nuclei from Subheading 3.3, **step 7** on top of the 300 μL of extraction buffer 3.
9. Centrifuge for 1 h at 16,000 RCF at 4 °C.
10. Remove the supernatant and resuspend the pellet in 310 μL of nuclei lysis buffer. Vigorously pipette the suspension 20 times and vortex for 20 s. Try to avoid forming bubbles, which can interfere with sonication. This step bursts the nuclei and releases high molecular weight chromatin into suspension.
11. Remove 10 μL of the lysed nuclei suspension and set aside on ice. This sample will serve as a pre-sonication quality control.
12. Give the tube with the remaining 300 μL suspension a brief pulse on the centrifuge in order to remove liquid from the lid. Gently resuspend any pellet via pipetting to ensure that all chromatin is in suspension. It is important that all of the liquid in the tube is on the bottom of the tube and is thus submerged in the sonicator water bath.
13. To fragment the chromatin, sonicate the chromatin suspension from Subheading 3.3, **step 12** using a Bioruptor. Set the Bioruptor to intervals on for 30 s, off for 30 s. Set the intensity level to maximum. Run for 15 min twice (for a total of 30 min). *See* **Note 7** for important details on Bioruptor usage.
14. Centrifuge sonicated chromatin at 16,000 RCF for 5 min at 4 °C. Transfer the supernatant into a new 1.5 mL Eppendorf tube and discard the pellet. The supernatant contains fragmented, soluble chromatin.
15. Briefly vortex the supernatant and remove 25 μL for the post-sonication quality control and for the ChIP input control.
16. Take the pre-sonication sample and 10 μL of the post-sonication sample (leaving 15 μL remaining for the ChIP input control) and bring each to 400 μL with ChIP elution buffer. Add 16 μL 5 M NaCl to each tube. Proceed with reverse cross-linking (Subheading 3.4, **step 8**) and the remaining steps as described in this protocol. Resuspend in 30 μL 10 mM Tris–HCl, pH 8.0 and run ~15 μL on a 1% agarose gel. Visualize

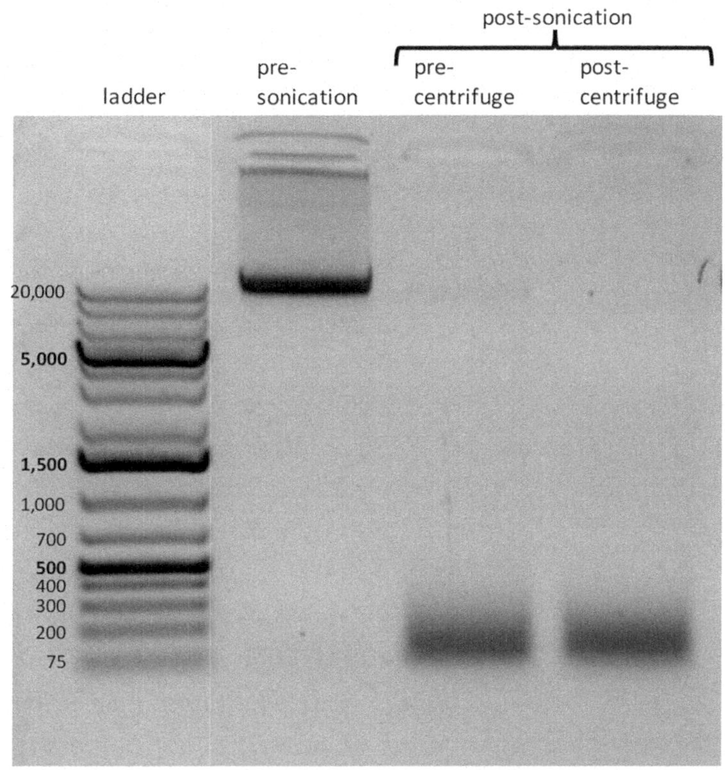

Fig. 2 Checking chromatin extraction and sonication quality with gel electrophoresis. A 1% agarose gel with the pre- and post-sonication samples, collected from Subheading 3.3, **steps 11** and **15**, respectively, and a post-sonication sample collected prior to centrifugation. This example shows successful chromatin extraction and sonication. The pre-sonication has a clear high molecular weight band of DNA. After sonication, that band nearly disappears. There is some visible residual high molecular weight DNA which was removed by centrifugation and thus no longer visible in the post-centrifuge sample. The sonicated fragment sizes range from ~150 to 400 bp

the gel to evaluate the success of the chromatin extraction and sonication (*see* Fig. 2 and **Note 8**).

17. With the remaining sonicated chromatin, either proceed to immunoprecipitation or store the chromatin at −80 °C. Chromatin can be stored at −80 °C for several months.

3.4 Chromatin Immunoprecipitation

All steps should be performed at 4 °C, unless indicated otherwise.

1. Dilute the remaining chromatin solution (~275 μL) tenfold with ChIP dilution buffer (containing protease inhibitor cocktail). This step lowers the SDS concentration to a suitable level, allowing for non-covalent interactions of chromatin with antibodies.

2. Wash the prepared antibody-coated beads 3 times with ChIP dilution buffer (containing water instead of protease inhibitor cocktail). The wash steps can be performed as previously described in Subheading 3.2, **step 2**. Do not vortex the beads from this point onward, since non-covalent interactions must be maintained. Instead, flick or invert the tubes. If tubes are inverted, ensure that beads are removed from the tube lids by inverting the tubes when placed on the magnetic rack.
3. Add 500 μL of the diluted chromatin solution from Subheading 3.4, **step 1** to the washed antibody-coated beads and incubate with rotation at 4 °C overnight.
4. Attach beads to magnet and remove supernatant.
5. Wash the chromatin-coated beads in the following manner:
   (a) 1 mL low salt wash buffer. Invert 10 times.
   (b) 1 mL high salt wash buffer. Invert 10 times.
   (c) 1 mL LiCl wash buffer. Invert 10 times.
   (d) 1 mL low salt wash buffer. Rotate 5 min.
   (e) 1 mL high salt wash buffer. Rotate 5 min.
   (f) 1 mL LiCl wash buffer. Rotate 5 min.
   (g) 1 mL TE buffer, invert 10 times. Remove the supernatant.
6. Add 200 μL of elution buffer to the beads. Vortex the beads for 10 s and incubate at 65 °C for 15 min. Place the tube on the magnetic rack and transfer the supernatant to a new 1.5 mL tube.
7. Repeat the elution from Subheading 3.4, **step 6**. Combine the two eluates for a total of 400 μL eluate.
8. Add 16 μL of 5 M NaCl to the eluate and reverse crosslink at 65 °C for 6 h to overnight. Reverse crosslink the 15 μL input sample from Subheading 3.3, **step 15** at the same time as the experimental samples. Bring the 15 μL input to 400 μL with chip elution buffer and proceed like with the other samples.
9. After reverse crosslinking, add 8 μL of 0.5 M EDTA, 16 μL Tris–HCl, pH 6.5, and 1 μL of 20 mC/mL Proteinase K Solution to the eluate and input. Incubate for 1 h at 50 °C.

3.5 DNA Purification

All steps should be performed at room temperature, unless indicated otherwise.

1. Add 200 μL of buffered phenol and 200 μL of chloroform to the proteinase-digested samples. Vortex for 25 s then centrifuge for 5 min at 12,000 RCF at room temperature.
2. Transfer the supernatant to a 1.5 mL tube. Add 100 μL 2.5 M potassium acetate, 800 μL of 100% Ethanol, and 2 μL of

20 mg/mL glycogen. Precipitate DNA at −20 °C for 2 h to overnight.

3. Centrifuge 30 min at maximum speed (preferably >20,000 RCF) at 4 °C. There should be a visible white pellet after centrifugation.
4. Remove the supernatant and wash the pellet with 1 mL of 70% ethanol.
5. Centrifuge the sample for 1 min at 12,000 RCF at room temperature.
6. Remove the supernatant and allow the pellet to air-dry (~20 min).
7. Resuspend the pellet in 50 μL of 10 mM Tris–HCl, pH 8.0. The solution can be heated to 50 °C for 10 min to facilitate DNA dissolution. This is the final ChIP DNA that can be used for Illumina library preparation or qPCR. *See* **Note 9** on expected ChIP DNA yields and purity, and **Note 10** and Fig. 3 on evaluating ChIP success.

3.6 DNA End-Repair

1. Transfer the dissolved ChIP DNA to a PCR tube suitable for a magnetic rack.
2. To the 50 μL of dissolved ChIP DNA, add the 22.85 μL of DNA end-repair master mix. Mix well by pipetting.
3. Incubate at room temperature for 45 min. This reaction repairs the sheared DNA ends and converts them to 5′-phosphorylated blunt ends.
4. Add 102 μL of Ampure beads directly to the tube (for a bead:sample ratio of 1.4:1, which size selects for ~150 bp fragments and larger).
5. Pipette up and down 10 times to mix, then incubate for 10 min at room temperature.
6. Place bead suspension on a magnetic rack until the solution is cleared. This can take several minutes.
7. Remove 100 μL of the supernatant, leaving ~10 μL behind to prevent bead loss.
8. Perform ethanol washes as follows:
   (a) Add 200 μL of 80% ethanol to the beads.
   (b) Move the tube to the opposite side of the magnet to pass the beads through the ethanol. Repeat this three times.
   (c) Remove and discard 190 μL of the supernatant.
   (d) Repeat for a second wash, adding another 200 μL 80% ethanol and moving the tube again.
   (e) Remove all supernatant and allow the tube to air-dry at room temperature for 5–10 min.

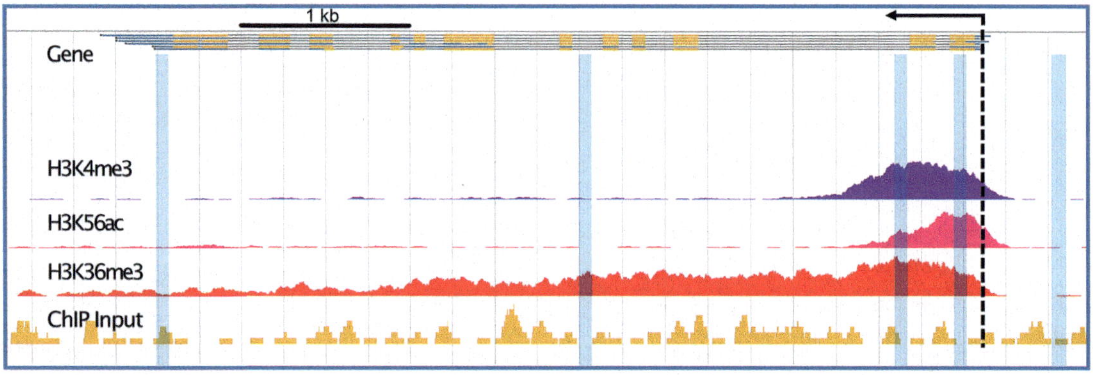

Fig. 3 Stereotypical histone mark positioning at transcribed genes. A genome browser screenshot of an actively transcribed maize gene. ChIP-seq on three histone modifications—trimethyl H3K4, trimethyl H3K36, and acetylated H3K56—displays the stereotypical patterns at transcribed genes. The gene model (shown at top) is transcribed to the left and the dashed line denotes the transcription start site. H3K4me3 and H3K56ac are typically enriched at the first couple of nucleosomes in the gene, while H3K36me3 is enriched throughout the transcription unit but tapers off toward the 3' end. The transparent blue boxes indicate potentially informative regions for qPCR enrichment analysis. In such regions, enrichment of ChIP signal over input should reveal the success of the ChIP

(f) Elute in 43 μL of 10 mM Tris–HCl, pH 8.0. Mix by pipetting to ensure full resuspension of beads.

(g) Incubate for 10 min at room temperature to elute the DNA.

3.7 DNA A-Tailing

1. To the tube containing beads and end-repaired DNA, add 5 μL NEBNext® dA-Tailing Reaction Buffer and 3 μL Klenow fragment (3′≥5′ exo-, 5 U/μL). Mix well by pipetting.
2. Incubate at 37 °C for 30 min.
3. Add 70 μL of new Ampure beads to a new 0.2 mL PCR tube.
4. Place the A-tailed tube on the magnetic rack until the solution is clear.
5. Transfer the entire supernatant (~50 μL) to the new tube containing the Ampure beads and mix well by pipetting. Discard the empty tube.
6. Incubate at room temperature for 10 min.
7. Place on magnetic rack until the solution is clear.
8. Repeat the ethanol washes 2× as described in Subheading 3.6, **step 8**. Air-dry the beads for 5–10 min.
9. Resuspend the beads in 22 μL of 10 mM Tris–HCl, pH 8.0 and elute the DNA for 10 min.

3.8 Ligate Illumina Sequencing Adapters to A-Tailed DNA Fragments

1. To the A-tailed DNA, add 2.5 μL T4 DNA ligase buffer, 1.25 μL T4 DNA ligase, and 1 μL of Illumina Truseq adapters. Mix well by pipetting.
2. Incubate at 16 °C for 16 h.
3. Add 50 μL of Ampure beads to a new 0.2 mL PCR tube.
4. Place the tube containing the ligated DNA on the magnetic rack and let the solution clear. Transfer the entire supernatant to the new tube containing 50 μL of bead suspension.
5. Resuspend by pipetting and incubate at room temperature for 10 min.
6. Clear the solution on the magnetic rack and repeat the 80% ethanol washes as described in Subheading 3.6, **step 8**. Air-dry the beads for 5–10 min.
7. Elute the beads into 20 μL of 10 mM Tris–HCl, pH 8.0.

3.9 PCR Amplification

1. Add the 30 μL of PCR master mix to the 20 μL of eluted adapter-ligated DNA (total 50 μL).
2. Run the following PCR program on the sample:
   - 95 °C for 2 min
   - 98 °C for 30 s
   - 15 cycles of:
     98 °C for 15 s
     60 °C for 30 s
     72 °C for 4 min
   - 72 °C for 10 min
   - 4 °C hold
3. After PCR amplification, perform Ampure DNA washes and 80% ethanol washes as detailed in Subheading 3.7, **step 3** (with 70 μL of Ampure beads) and Subheading 3.6, **step 8**, respectively.
4. The amplified DNA sample is ready for Illumina sequencing.

4 Notes

1. Preparation of stock solutions
   The ChIP dilution buffer, extraction buffer 1, extraction buffer 2, extraction buffer 3, and nuclei lysis buffers should be prepared as stock solutions with PMSF, BME, and protease inhibitor cocktail omitted. Add these reagents before use. The cOmplete™ Mini EDTA-free Protease Inhibitor Cocktail comes as solid tablets. These tablets can be dissolved in ultra-pure water as a 10× concentrate and stored at 4 °C for 3 weeks

or at −20 °C indefinitely. The 10× protease cocktail should be diluted to 1× in the working solution. Filter-sterilize the stock solutions and store them at 4 °C.

2. Input tissue quantity

   This protocol is flexible regarding the amount of input chromatin. However, more chromatin is not necessarily better. We have achieved high-quality results with less than 0.5 million maize nuclei. This number can be pushed even lower with a high-quality nuclei prep that is largely free of debris. Cleaner nuclei result in cleaner ChIP signal. Plant tissues with higher nuclei count per weight (and thus less debris per unit nuclei) tend to yield higher quality ChIP results. For example, highly proliferative tissues, such as maize inflorescence primordia or the basal portion of young leaves, give better results than do highly elongated tissues, such as mature leaves. In our experience, we have never needed to exceed 1 g of tissue fresh weight in order to acquire sufficient chromatin for histone modification ChIP.

3. Choosing an antibody

   Because the amino acid sequences of histones (particularly H3) are highly conserved among eukaryotes, antibodies that recognize specific histone modifications in animal and fungal species are commercially available and can be used to study histone modifications in plants. Choosing antibodies for ChIP is one of the most important considerations when designing an experiment. Some antibodies show more binding promiscuity than do others. Antibodies targeting acetylated lysines in the N-terminal tail of histone H3 tend to bind promiscuously. Furthermore, antibodies vary in their signal-to-noise ratios. Polyclonal antibodies can vary between lots as well. Therefore, it is advisable to test the specificity of commercially available antibodies against synthetic modified histones in peptide competition assays. Once a high-quality antibody lot has been identified, the same lot should be used for all experiments. The catalog numbers and lot numbers of antibodies should be recorded and included in publications in order to facilitate reproducibility. Some commonly used antibodies with which we have found success are listed in Table 1.

4. Procedural time considerations

   The entire ChIP procedure can comfortably be completed within 4 days. If you are performing ChIP on a species and tissue that has already been experimentally optimized: On day 1, perform crosslinking, perform the chromatin isolation, and incubate the antibody-coated beads with the sonicated chromatin overnight. On day 2, wash the beads and reverse crosslink overnight. On day 3, finish the ChIP protocol and purify the DNA. Run the agarose gel to make sure that the sonication worked. Perform the DNA end-repair, A-tailing, and ligate adapters overnight. On day 4, perform PCR. This timeline assumes that you already know the right sonication conditions for the species or tissue type that you are working with. If working with a plant species that you have not worked with before, it is recommended that you check the quality of the chromatin extraction and sonication (via agarose gel) prior to proceeding with immunoprecipitation. This adds an additional day to the procedure because the pre- and post-sonication DNA samples must be reverse crosslinked and purified.

5. Formaldehyde and PMSF usage

   Add formaldehyde inside of a fume hood. Add PMSF just before use as it is unstable in aqueous solution. Ensure that the PMSF has been fully dissolved in the isopropanol prior to pipetting (0.1 M PMSF precipitates in isopropanol at −20 °C). Vortex the solution to ensure that it is fully mixed.

6. Precautions during vacuum infiltration

   After applying the vacuum to the submerged tissue, gas bubbles exiting the tissue should be apparent. The goal is to replace the gas inside of the tissue with crosslinking solution to ensure that the high surface area within the tissue (especially leaves) interior contacts the solution. Watch the tube to make sure that the dim sum bag does not float to the surface when applying the vacuum. If this occurs, release the vacuum, rearrange the twist tie, and reapply the vacuum.

7. Bioruptor usage details

   The depth of the Bioruptor water bath should be ~5 cm and the water should be free of debris. The bath should be chilled to 4 °C. If available, use a pump to keep the water cool during sonication. If a pump is unavailable, add ice to the bath every 15 min. Without active cooling, the sonic energy will heat up the water bath. The Bioruptor is designed to work on 300 μl volumes in 1.5 mL Eppendorf tubes. The shearing force from the sound waves fragments the chromatin. Linker DNA is preferentially sheared, leaving intact nucleosomes. Thirty minutes of sonication ensures that the size of chromatin fragments approaches single-nucleosome resolution.

8. Visualizing chromatin extraction and sonication success

   The success of the ChIP protocol is contingent on the isolation of high-quality chromatin. Therefore, it is advisable to evaluate the success of the chromatin extraction and fragmentation prior to proceeding with immunoprecipitation. Generally, if the chromatin has fragmented to a size range of ~150–500 bp, the ChIP should result in high signal to background (assuming the use of a high-quality antibody). Smaller chromatin fragments allow for higher resolution profiles of histone modifications. Residual large chromatin fragments can increase background noise. Figure 2 demonstrates successful chromatin extraction and sonication. The pre-sonication sample contains exclusively high molecular weight chromatin, which is absent in the post-sonication sample.

9. Expected ChIP DNA yields and purity

   Unamplified DNA yields from histone ChIP can vary substantially. Concentrations are often <10 ng/μl and thus cannot be accurately quantified with a nanodrop spectrophoto-meter. The concentration of input DNA, when dissolved in 50 μl, typically exceeds 50 ng/μl. A slight peak at 270 nm on the spectrophotometer indicates phenol carryover. Increased absorbance below 230 nm is attributable to the added glycogen. This is normal and does not appear to interfere with subsequent qPCR or Illumina library preparation.

10. Evaluating ChIP success

    Many histone covalent modifications show stereotypical enrichment patterns at genes, depending on the genes' expression statuses. These stereotypical patterns tend to be conserved across eukaryotes, and thus enrichment of these patterns can serve as positive controls to evaluate the efficacy of the ChIP. Such evaluations can be performed with qPCR prior to sequencing or after sequencing by visualizing in a genome browser. However, it is advisable that, prior to sequencing, qPCR is used to ensure that ChIP enrichment worked as expected in order to avoid sequencing poor samples. Figure 3 shows an example of H3K4me3, H3K36me3, and H3K56ac ChIP-seq results from maize leaves. This example shows the stereotypical patterns of enrichment at expressed genes and how primers can be used to detect enrichment at this locus with qPCR.

Acknowledgments

This work was supported by NSF PGRP grant 1546867 to X.Z.

References

1. Ji X, Dadon DB, Abraham BJ et al (2015) Chromatin proteomic profiling reveals novel proteins associated with histone-marked genomic regions. Proc Natl Acad Sci U S A 112:3841–3846
2. Kouzarides T (2007) Chromatin modifications and their function. Cell 128:693–705
3. Li B, Carey M, Workman JL (2007) The role of chromatin during transcription. Cell 128:707–719
4. Martin C, Zhang Y (2005) The diverse functions of histone lysine methylation. Nat Rev Mol Cell Biol 6:838–849
5. Albert I, Mavrich TN, Tomsho LP et al (2007) Translational and rotational settings of H2A.Z nucleosomes across the Saccharomyces cerevisiae genome. Nature 446:572–576
6. Barski A, Cuddapah S, Cui K et al (2007) High-resolution profiling of histone methylations in the human genome. Cell 129:823–837
7. Johnson DS, Mortazavi A, Myers RM et al (2007) Genome-wide mapping of in vivo protein-DNA interactions. Science 316:1497–1502
8. Mikkelsen TS, Ku M, Jaffe DB et al (2007) Genome-wide maps of chromatin state in pluripotent and lineage-committed cells. Nature 448:553–560

Chapter 10

Whole-Genome Bisulfite Sequencing and Epigenetic Variation in Cereal Methylomes

Taiji Kawakatsu

Abstract

Whole-genome bisulfite sequencing (WGBS) is a technique used for the analysis of genome-wide DNA methylation patterns (DNA methylomes) at a single-base resolution. Here, I describe a simple DNA extraction method from rice endosperm and the universal protocol of WGBS, MethylC-sequencing library preparation. The use of benzyl chloride allows for the extraction of high-quality genomic DNA from starchy endosperm, while sodium bisulfite converts unmethylated cytosine to uracil, whereas methylated cytosine is unchanged. The bisulfite conversion of whole genome sequencing libraries before the final amplification step allows for the discrimination of methylated from unmethylated cytosines in a genome-wide manner.

Key words DNA methylomes, Rice, Cereal, Endosperm, Whole-genome bisulfite sequencing, MethylC-sequencing

1 Introduction

DNA methylation is an epigenetic mark occurring at cytosine residues, which has impacts on gene regulation associated with development, pluripotency, cancer, imprinting, and X-chromosome inactivation [1]. Cytosines are methylated in three distinct contexts: CG, CHG, and CHH types, where H = A, T, or C [1]. DNA methylation only in the CG context within gene bodies is associated with constitutive gene expression, whereas DNA methylation in all three contexts is associated with transcriptional inactivation by RNA-directed DNA methylation [2]. Therefore, information about the methylation context can reveal the hidden biological function of DNA methylation.

There are two main types of methods to analyze genome-wide DNA methylation patterns (methylomes): (1) bisulfite conversion-based methods, and (2) methylated DNA enrichment-based methods [3]. Treatment of DNA with sodium bisulfite converts unmethylated cytosine to uracil, which is converted to thymine

Luis M. Vaschetto (ed.), *Cereal Genomics: Methods and Protocols*, Methods in Molecular Biology, vol. 2072, https://doi.org/10.1007/978-1-4939-9865-4_10,

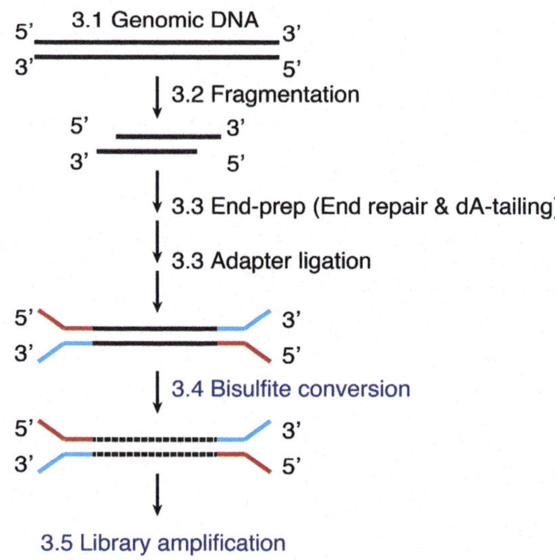

Fig. 1 Schematic of MethylC-seq library preparation procedures. Genomic DNA is fragmented, followed by end-prep, adapter ligation, bisulfite conversion, and library amplification steps

during subsequent PCR. Whole genome bisulfite sequencing (WGBS) has been developed by applying bisulfite conversion to whole genome sequencing before library amplification (Fig. 1) [4]. WGBS is widely adopted for profiling methylomes because it can determine the DNA methylation status at 90–95% of cytosine positions in the genome at a single-base resolution [5].

Cereal endosperms are a major source of calories worldwide. Endosperms have distinct DNA methylomes compared with other tissues [6], and imprinting caused by DNA methylation in the endosperm is often associated with hybrid incompatibility [7, 8]. Therefore, the demand for methylome analysis of cereal endosperm has been increasing.

Here, I demonstrate highly efficient DNA extraction methods from starchy rice endosperm and WGBS library preparation methods. Both procedures can be performed in 2 days. I utilize simplified versions of the previously described benzyl chloride-based DNA extraction method and the MethylC-sequencing (MethylC-seq) library preparation method [5, 9].

Benzyl chloride is an organic compound that is mostly insoluble in water. It destroys cell walls by reacting with the hydroxy group in polysaccharides. It also extracts starch as well as proteins and cell debris from the aqueous phase that includes nucleic acids. This enables the extraction of high-quality genomic DNA from starch endosperm that cannot otherwise be achieved by column-based kits such as the Qiagen DNeasy plant mini kit. Genomic DNA extracted using benzyl chloride can undergo MethylC-seq library preparation as well as restriction enzyme digestion without any further purification (Figs. 2 and 3).

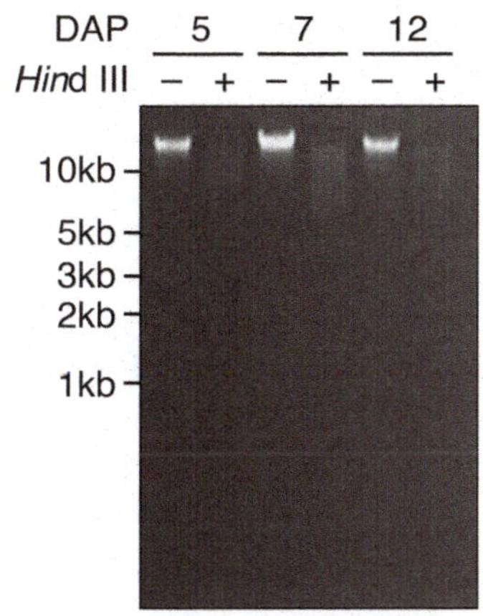

Fig. 2 Genomic DNA extracted from immature rice endosperm. Genomic DNA was extracted from developing endosperms at 5, 7, and 12 days after pollination (DAP). DNA extracted using the method described in this chapter is much larger than 10 kb (–), whereas HindIII cleavage for 15 min (+) causes smearing, showing that the extracted DNA is of good quality

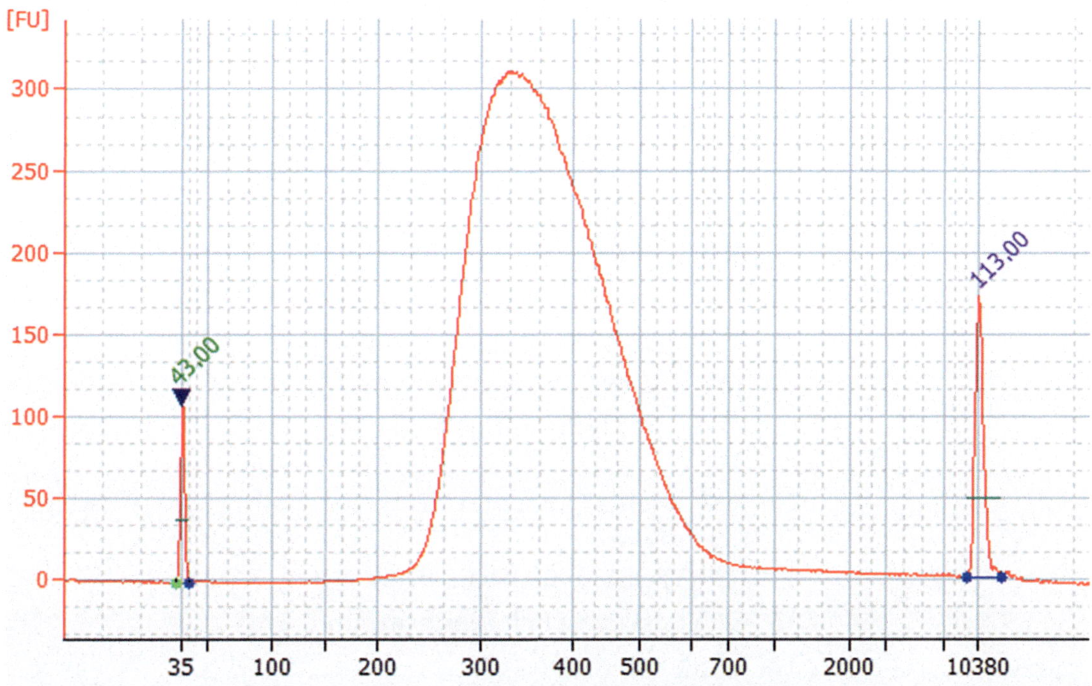

Fig. 3 Size distribution of MethylC-seq library prepared by the method described in this chapter. Most DNA fragments in the library are approximately 250–600 bp, so insert sizes are 120–470 bp, which is almost equivalent to the range of double-size selected DNA in Subheading 3.2

2 Materials

2.1 DNA Extraction

1. DNA extraction buffer: 100 mM Tris–HCl, 40 mM EDTA, 2% SDS, pH 9.0. Autoclave and store at room temperature.
2. Benzyl chloride.
3. Precipitation buffer: 3 M sodium acetate. Autoclave and store at room temperature.
4. Isopropyl alcohol.
5. 70% ethanol.
6. Elution buffer: 10 mM Tris–HCl, pH 8.5.
7. RNase A solution: 100 mg/mL RNase A.
8. Qubit dsDNA High Sensitivity kit (Thermo Fisher).
9. Qubit fluorometer (Thermo Fisher).
10. Multibeads shocker tube (Yasui Kikai).
11. Multibeads shocker (Yasui Kikai).

2.2 MethylC-Seq Library Preparation

1. microTUBE AFA Fiber Pre-Slit Snap-Cap (Covaris).
2. Covaris Focused-ultrasonicator M220.
3. 1.5 mL DNA LoBind tubes (Eppendorf).
4. DynaMag-2 magnet (Thermo Fisher).
5. Agencourt AMPure XP (Beckman) (*see* **Note 1**).
6. Elution buffer: 10 mM Tris–HCl, pH 8.5
7. 0.2 mL PCR tubes.
8. PCR machine.
9. NEBNext Ultra II End-prep module (New England Biolabs).
10. NEBNext Ultra II Ligation module (New England Biolabs).
11. 80% ethanol.
12. EZ DNA Methylation-Gold kit (Zymo Research) (*see* **Note 2**).
13. KAPA HiFi HotStart Uracil+ ReadyMix (Kapa Biosystems) (*see* **Note 3**).
14. NEXTflex Bisulfite-Seq Barcodes (Bioo Scientific).
15. Qubit dsDNA High Sensitivity kit.
16. Qubit Fluorometer.
17. Agilent Bioanalyzer 2100.
18. Agilent DNA High Sensitivity kit.

3 Methods

3.1 DNA Extraction from Endosperm

1. Snap-freeze three immature endosperms in a 2 mL Multibeads shocker tube with liquid nitrogen. Grind to a fine powder using Multibeads shocker (2000 rpm, 7 s) (*see* **Notes 4** and **5**).
2. Add 300 μL of DNA extraction buffer and vortex for 5 s.
3. Add 150 μL of benzyl chloride and vortex for 5 s (*see* **Note 6**).
4. Incubate tubes at 50 °C for 10 min, with occasional vortexing.
5. Add 150 μL of precipitation buffer and vortex for 5 s.
6. Place tubes on ice for 15 min.
7. Centrifuge at >12,000 *g* for 10 min at 4 °C.
8. Collect supernatant in a new tube.
9. Add 1× volume of isopropyl alcohol and mix.
10. Place tubes on ice for 10 min.
11. Centrifuge at >12,000 *g* 10 min at 4 °C.
12. Discard supernatant and add 1 mL 70% ethanol.
13. Centrifuge at >12,000 *g* for 5 min at 4 °C.
14. Discard supernatant and air-dry for 5 min.
15. Dissolve in 100 μL elution buffer (*see* **Note** 7).
16. Add 1 μL of RNase A solution and incubate at 37 °C for 10 min.
17. Add 100 μL of isopropyl alcohol and 10 μL of precipitation buffer.
18. Repeat **steps 11–15** (*see* **Note 8**).
19. Add 100 μL of elution buffer.
20. Measure DNA concentration with Qubit dsDNA High Sensitivity kit and Qubit fluorometer (*see* **Note 9**).

3.2 DNA Fragmentation

1. Transfer 2 μg of genomic DNA in 130 μL of elution buffer to a microTUBE AFA Fiber Pre-Slit Snap-Cap.
2. Run Covaris with 200 bp target mode.
3. Collect 100 μL of sonicated DNA in a DNA LoBind tube.
4. Add 60 μL of Agencourt AMPure XP beads and mix well by pipetting or vortexing.
5. Incubate at room temperature for 10 min.
6. Place tubes on DynaMag-2 magnet for 5 min.
7. Collect the supernatant (containing DNA <500 bp) into a new DNA LoBind tube. Discard the tube with the bead pellet (containing DNA >500 bp).

8. Add 80 μL of Agencourt AMPure XP to the upper size selected supernatant and mix well by pipetting or vortexing.
9. Incubate at room temperature for 10 min.
10. Place tubes on DynaMag-2 magnet for 5 min.
11. Discard the supernatant containing DNA <150 bp.
12. Add 500 μL of 80% ethanol to the tube and wait 30 s.
13. Discard the supernatant. Add 500 μL of 80% ethanol to the tube and wait 30 s again.
14. Discard the supernatant. Try to remove any residual ethanol.
15. Keep the lid open and air-dry at room temperature for 3 min or until the bead pellet dries (*see* **Note 10**).
16. Add 26 μL of elution buffer and mix well by pipetting.
17. Incubate at room temperature for 2 min.
18. Place tubes on DynaMag-2 magnet for 2 min.
19. Collect the supernatant into a PCR tube.

3.3 End-prep and Adaptor Ligation (*See* Note 11)

1. Add 3.5 μL of End-prep buffer and 1.5 μL of End-prep enzyme, included in the NEBNext Ultra II End-prep module, to the double size selected DNA.
2. Perform the following end-prep reaction in the PCR machine:

End-prep reaction program	
20 °C	30 min
65 °C	30 min
4 °C	Hold

3. Add 15 μL of ligation master mix, 0.5 μL of ligation enhancer (included in the NEBNext Ultra II Ligation module), and 1.25 μL of NEXTflex Bisulfite-Seq Barcodes.
4. Perform the following adaptor ligation reaction in the PCR machine:

Adapter ligation reaction program	
20 °C	15 min
4 °C	Hold

5. Collect the adaptor ligated DNA into a new DNA LoBind tube.
6. Perform Subheading 3.2, **steps 8–18** with 38 μL of Agencourt AMPure XP and elute purified products with 50 μL of elution buffer.
7. Collect the supernatant into a new DNA LoBind tube.

8. Perform Subheading 3.2, **steps 8–18** with 50 μL of AMPure XP beads and elute with 21 μL of elution buffer.
9. Collect the 20 μL of supernatant into a new PCR tube.

3.4 Bisulfite Conversion

1. Add 900 μL of water, 300 μL of M-dilution buffer, and 50 μL of M-dissolving buffer into a tube of CT conversion reagent, included in the EZ DNA Methylation-Gold kit. Vigorously vortex for 10 min at room temperature.
2. Add 130 μL of CT conversion reagent to adaptor-ligated DNA fragments from Subheading 3.3, **step 9**.
3. Perform the following bisulfite conversion reaction in the PCR machine:

Bisulfite conversion reaction program	
98 °C	10 min
64 °C	150 min
4 °C	Hold (up to 20 h)

4. Place a Zymo-Spin IC Column into a collection tube supplied in the EZ Methylation-Gold kit. Add 600 μL of M-binding buffer to a Zymo-Spin IC Column.
5. Add the sample from Subheading 3.4, **step 4** to a Zymo-Spin IC Column with M-binding buffer. Close the lid and mix well by inverting the column several times.
6. Centrifuge at >12,000 *g* for 30 s at room temperature. Discard the flow-through and place the column into the same tube.
7. Add 100 μL of M-wash buffer to the column and centrifuge at >12,000 *g* for 30 s at room temperature.
8. Add 200 μL of M-desulfonation buffer to the column and incubate at room temperature for 15–20 min.
9. Centrifuge at >12,000 *g* 30 s at room temperature.
10. Add 200 μL of M-wash buffer to the column. Centrifuge at >12,000 *g* for 30 s at room temperature.
11. Repeat Subheading 3.4, **step 10**.
12. Discard the flow-through and place the column into the same tube. Centrifuge at >12,000 *g* for 2 min at room temperature.
13. Add 24 μL of M-elution buffer to the column. Centrifuge at >12,000 *g* for 30 sec at room temperature and elute the bisulfite converted DNA.

3.5 Library Amplification

1. Add 25 μL of KAPA HiFi Uracil+ Ready Mix and 2.5 μL of primers supplied with NEXTflex Bisulfite-Seq Barcodes.

2. Perform the following library amplification in the PCR machine (*see* **Note 12**):

Library amplification program	
95 °C	2 min
98 °C	30 s
4 cycles of	
98 °C	10 s
60 °C	30 s
72 °C	30 s
4 °C	Hold

3. Collect the reaction mix into a new DNA LoBind tube. Perform Subheading 3.2, **steps 8–18** with 50 μL of Agencourt AMPure XP and elute with 21 μL of elution buffer.
4. Collect 20 μL of supernatant into a new DNA LoBind tube.
5. Measure the DNA concentration with a Qubit dsDNA High Sensitivity kit and Qubit fluorometer.

4 Notes

1. SPRI beads from other suppliers work. However, the amount of beads should be optimized because slight changes in the PEG8000 concentration can greatly affect the size of bead-bound DNA.
2. I recommend the EZ DNA Methylation-Gold kit or MethylCode Bisulfite Conversion kit. I encountered mapping bias when using the EpiTect Bisulfite Conversion kit.
3. The KAPA HiFi HotStart Uracil+ ReadyMix can amplify libraries with less mapping bias and higher fidelity.
4. This DNA extraction protocol works with other plant tissues below 100 mg, such as leaves and roots.
5. Multibeads shocker can be substituted with similar equipment, such as TissueLyser and Shake Master. Pulverization can be performed with a mortar and pestle.
6. Benzyl chloride is hazardous. Use a fume hood to avoid exposure to benzyl chloride and its vapor.
7. Incubate at 50 °C for 10 min if the genomic DNA is difficult to dissolve.

Table 1
PCR cycles required for different amounts of input DNA

Input DNA	PCR cycles
2 μg	4
1 μg	5
500 ng	6
250 ng	7
100 ng	8
50 ng	9
25 ng	10
10 ng	12
5 ng	13
1 ng	15

8. Subheading 3.1, **steps 17** and **18** are optional but highly recommended because they improve the DNA quality (in terms of 260/280 and 260/230 optical densities).
9. Use Qubit or another fluorescence-based DNA quantification method because UV spectrophotometers, such as NanoDrop, can overestimate the DNA quantity through contamination.
10. Do not over-dry the beads longer than 5 min because this will greatly affect the recovery rate.
11. The reaction volume used in this technique is half that of the manufacturer's protocol.
12. This protocol works for limited amounts of DNA as low as 1 ng for *Arabidopsis*, whose genome size is 119 Mb. The lowest input should be higher for species with larger genomes (e.g., 3 ng for rice with a genome size of 370 Mb). Amplification cycles should be increased for lower inputs, as indicated in Table 1. However, extra amplification cycles increase redundancy, so more sequencing may be required.

Acknowledgments

This work is supported by JSPS KAKENHI (17H05851 and 19H04873).

References

1. Law JA, Jacobsen SE (2010) Establishing, maintaining and modifying DNA methylation patterns in plants and animals. Nat Rev Genet 11:204–220
2. Kawashima T, Berger F (2014) Epigenetic reprogramming in plant sexual reproduction. Nat Rev Genet 15:613–624
3. Lister R, Ecker JR (2009) Finding the fifth base: genome-wide sequencing of cytosine methylation. Genome Res 19:959–966
4. Lister R, O'Malley RC, Tonti-Filippini J et al (2008) Highly integrated single-base resolution maps of the epigenome in Arabidopsis. Cell 133:523–536
5. Urich MA, Nery JR, Lister R et al (2015) MethylC-seq library preparation for base-resolution whole-genome bisulfite sequencing. Nat Protoc 10:475–483
6. Zemach A, Kim MY, Silva P et al (2010) Local DNA hypomethylation activates genes in rice endosperm. Proc Natl Acad Sci U S A 107:18729–18734
7. Lafon-Placette C, Hatorangan MR, Steige KA et al (2018) Paternally expressed imprinted genes associate with hybridization barriers in Capsella. Nat Plants 4:352–357
8. Kinoshita T (2018) A parental tug-of-war. Nat Plants 4:329–330
9. Zhu H, Qu F, Zhu LH (1993) Isolation of genomic DNAs from plants, fungi and bacteria using benzyl chloride. Nucleic Acids Res 21:5279–5280

Chapter 11

Genome-Wide Identification of Allele-Specific Gene Expression in a Parent-of-Origin Specific Manner

Chen Chen and Kevin Begcy

Abstract

Upon fertilization, normal endosperm and embryo development require the contribution of both the maternal and paternal genomes. However, certain genes are expressed in a parent-of-origin-dependent manner, an epigenetic phenomenon known as genomic imprinting. Despite the blast of new technologies and the crucial advances of the past decades in the epigenetics field, novel imprinted genes are yet to be discovered and thus key regulators of early seed development. Using rice plant as a model, we describe a method for the identification of imprinted genes based on an RNA-Seq approach, which allows the identification of maternal and paternal gene expression in a parent-of-origin-specific manner.

Key words Cereals, Imprinting, Allele-specific gene expression, Epigenetics, Rice, Endosperm, Embryo

1 Introduction

In angiosperms, the vast majority of the genome is expressed equally from maternal and paternal alleles. However, some genes present a biased expression of alleles that depends upon the parent of origin, an epigenetic phenomenon referred to as imprinting. Histone modifications or DNA methylation or a combination of both mechanisms are key epigenetic regulators of genomic imprinting [1–4]. Genes that preferentially or exclusively express one allele of their parents are termed maternally expressed genes (MEG) or paternally expressed genes (PEG) [2]. Despite a widespread interest in genomic imprinting and the identification of hundreds of putative imprinted loci, there is a slight conservation of imprinted genes among crop species, suggesting a dynamic and rapid evolution of genomic imprinting.

Genome-wide identification of imprinted genes includes reciprocal crosses of closely related plant species, harvesting of developing endosperm, construction of libraries, RNA sequencing, alignment and single-nucleotide polymorphism (SNP) calling,

Luis M. Vaschetto (ed.), *Cereal Genomics: Methods and Protocols*, Methods in Molecular Biology, vol. 2072,
https://doi.org/10.1007/978-1-4939-9865-4_11,

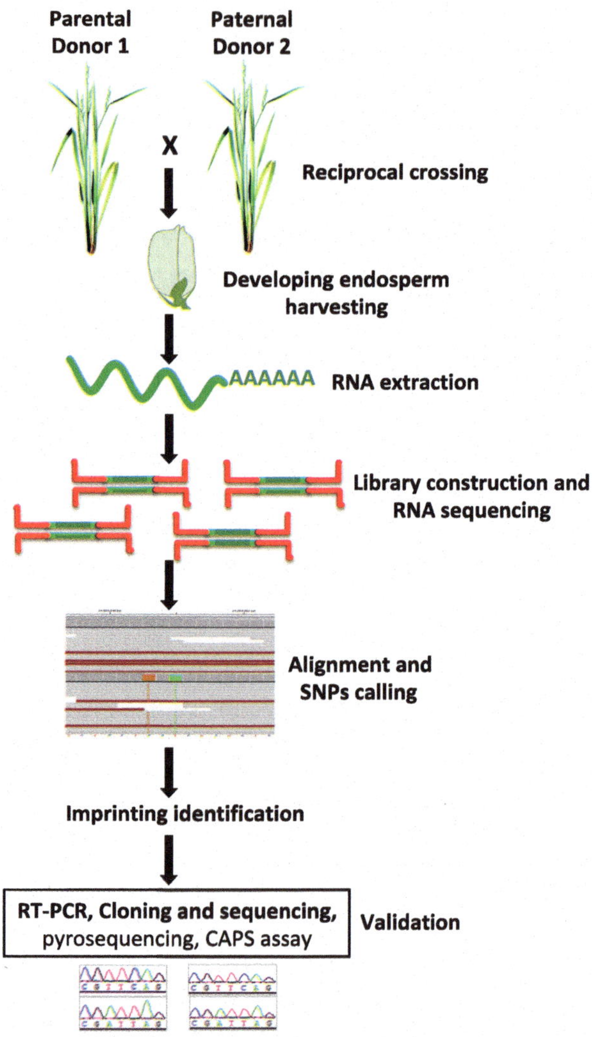

Fig. 1 Schematic illustration of major steps involved in genome-wide identification of allele-specific gene expression in a parent-of-origin-specific manner

identification of allele-specific expression (imprinting), and validation (Fig. 1). One of the main challenges in the identification of imprinted genes is the variation in the allele-specific expression attributed to possible expression of one or several isoforms, tissue-specific expression or developmental regulation [5–7]. Despite the blast of new technologies and the crucial advances in the past decades, the elucidation of mechanisms and biological functions of genomic imprinting is not well understood, and therefore the accurate identification of the whole set of imprinted genes present in cereal crop genomes is crucially needed. Several approaches have been used to identify genomic imprinting, including screenings based on gene expression analysis [2, 8], epigenetic features and differential methylation [9, 10]. Notably, imprinting is predominantly expressed in the endosperm of plants, though few imprinted genes have been found in the embryo but with controversy. Using rice as a model, we described a detailed step-by-step protocol to identify allele-specific gene expression in a parent-of-origin-specific manner in cereals.

2 Materials

2.1 Plant Material

Indica and *japonica* belong to different subspecies of the Asian cultivated rice (*Oryza sativa*), showing a relative distant genetic relationship [11]. Hybrid seeds derived from intraspecific crosses between *indica* and *japonica* are usually viable. More importantly, the abundant DNA polymorphisms present across *japonica* and *indica* genomes are key features to distinguish parental alleles of most of the genes encoded by the rice genome. Therefore, here we used *O. sativa* ssp. *indica* and *O. sativa* ssp. *japonica* as parental species to perform reciprocal crosses for the identification of imprinted genes in rice (*see* **Note 1**). Similar flowering time between the parents is a desirable trait (*see* **Note 2**).

2.2 Reagents

1. TRIzol Reagent (Life Technologies) or TRI reagent or similar.
2. DEPC (RNase-free) water (*see* **Note 3**).
3. Chloroform (Fisher) or similar.
4. Isopropyl alcohol (2-propanol) (Sigma-Aldrich) or similar.
5. Ethanol 75% (in DEPC-treated water).
6. KAPA SYBR® FAST qPCR Master Mix (2×) Universal Kit (Sigma-Aldrich) or similar.
7. TruSeq RNA Library Prep Kit v2 (Illumina) or similar.
8. pGEM®-T Easy Vector (Promega) or similar TA cloning vector.
9. QIAprep® Miniprep (Qiagen) or similar.
10. LB medium or similar.
11. GoTaq™ DNA Polymerase (Promega) or similar.
12. T4 DNA ligase (Promega) or similar.
13. Dynabeads™ Oligo(dT)25 (ThermoFisher).
14. RNAlater™ Stabilization Solution (ThermoFisher) or similar.
15. RNeasy Plant Mini Kit (Qiagen) or similar.

2.3 Equipment

1. Tissue Lyser II (Qiagen) or a pestle and mortar.
2. Centrifuge 5424R (Eppendorf) or similar.
3. Microcentrifuge 5702 (Eppendorf) or similar.
4. MasterCycler RealPlex (Eppendorf) or similar.

5. MaxQ 420 High Performance Shaker (ThermoFisher) or similar.
6. Agilent 2100 Bioanalyzer (Agilent Technologies) or similar.
7. cBot 2 System (Illumina) or similar.
8. Qubit® 2.0 Fluorometer (Life Technologies) or similar.
9. Thermo Scientific Precision GP 20 Water Bath or similar.

3 Methods

3.1 Reciprocal Crossing of Rice

1. Germinate rice seeds (*O. sativa* ssp. *indica* and *O. sativa ssp. japonica* or any cultivar, inbred line or variety of interest) in water under dark conditions. After seed germination, transplant seedlings to soil and grow them under controlled conditions (27–30 °C) in the greenhouse until approaching anthesis [2, 12–14].
2. Emasculate every spikelet of a panicle and carefully remove the anthers completely at the day before flowering of each of the plants that are going to be used as maternal parent. Use pollinating bags to cover emasculated panicles.
3. As soon as anthers of the paternal parent begin to extrude, shed pollen grains into the emasculated spikelet. Alternatively, you can use three-line hybrid rice strains to construct reciprocal crosses (*see* **Note 4**). A cytoplasmic male sterile (CMS) line, which fails to produce viable pollens usually due to a toxic sterility gene encoded by the mitochondria genome, can be used as the maternal parent. The corresponding maintainer line of a CMS line has the same nuclear genome but lacks the cytoplasmic sterililty gene. The maintainer line is fertile; therefore, it can be used as pollen donor for the reciprocal crosses. Because there is no need to remove the anthers of an emasculated CMS spikelet, it is timesaving and more efficient for crossing. Moreover, it greatly reduces the false hybridization events when using the CMS lines and maintainer lines for cross-pollination.
4. Since imprinting is a highly dynamic process, collecting developing seeds at different days after pollination (DAP) would allow to distinguish changes throughout plant development.
5. It is important to separate endosperm (3n) from embryo (2n) and maternal tissues (2n), since they have a different genomic architecture and thus show genes with different parent-of-origin-dependent expression pattern.
6. For caryopsis at 5 days after fertilization (DAF) or later, separation of the tissues is not as difficult and laborious as during early stage samples. However, caryopsis at early developmental

stages is difficult to separate from the endosperm, embryo and maternal tissues. Laser capture microdissection (LCM) technology can be applied if one wants to discover the imprinted genes for the very-early-developed endosperm.

7. For milky endosperm collection (seeds collected 7 DAF), a pipette of 10 μL can be used to pull out all the endosperm contained in each seed, thus, avoiding maternal tissue contamination.
8. For endosperm at dough stage (seeds collected 12–15 DAF), remove carefully seed coat using fine micro dissector forceps.
9. Collected samples need to be frozen in liquid nitrogen or submerged immediately in a 1.5 or 2 mL Eppendorf tube containing RNAlater stabilization solution.

3.2 RNA Extraction

RNA TRIzol isolation works fine for the early endosperm samples. However, for samples older than 7 DAF, which contain high levels of starch, some modifications need to be made in the isolation protocol (*see* **Note 5**).

3.2.1 Homogenization

1. Transfer plant material to a Tissue Lyser II adapter prechilled in liquid nitrogen and homogenize with a frequency of 30 Hz for 1 min. Alternatively, plant material can be ground using a pestle and mortar to a fine powder.
2. From the fine powder material transfer 50–100 mg to an RNase-free 1.5 mL microcentrifuge tube (*see* **Note 6**).
3. Add 1 mL of TRIzol to every 50–100 mg of tissue.
4. Vortex samples vigorously for 30 s.
5. Incubate samples for 15 min at room temperature.
6. Carefully add 0.2 mL of chloroform to each tube and vortex vigorously for 15 s.
7. Incubate samples for 5 min at room temperature.
8. Centrifuge samples for 15 min at 12,000 × *g* at 4 °C (*see* **Note 7**).
9. Transfer the colorless upper aqueous phase to a new RNase-free 1.5 mL microcentrifuge tube (*see* **Note 8**).
10. Add 0.5 mL of isopropyl alcohol to precipitate RNA.
11. Mix gently all sample mixtures by inverting them 10 times (*see* **Note 9**).
12. Incubate samples at room temperature for 5 min.
13. Centrifuge at maximum speed for 10 min at 4 °C.
14. Transfer the sample into a new tube.
15. Incubate for at least for 1 h at −20 °C.
16. Centrifuge for 10 min at 12,000 × *g* at 4 °C (*see* **Note 10**).

3.2.2 RNA Wash

1. Discard the supernatant.
2. Wash the RNA pellet with 1 mL of prechilled 75% ethanol.
3. Mix sample by vortexing (*see* **Note 11**).
4. Centrifuge at 12,000 × *g* for 5 min at 4 °C.
5. Remove supernatant.
6. Let the pellet air-dry for 10 min.
7. Dissolve the pellet in 30–60 μL of RNase-free water.

3.3 Library Construction and RNAseq

1. Library construction largely depends on the kit used for its synthesis. For paired-end libraries we recommend TruSeq® RNA Sample Preparation Kit. A detailed protocol can be found enclosed in the TruSeq® RNA Sample Preparation Guide.
2. Purification of Poly-A-containing mRNA molecules can be performed using poly-T oligo-attached magnetic beads or similar.
3. For library quality control, purified libraries can be quantified by Qubit® 2.0 Fluorometer or similar.
4. For validation, confirmation of insertion size and concentration, an Agilent 2100 Bioanalyzer or similar can be used.
5. Cluster was generated by cBot 2 system or similar, with a concentration of 10 pM per library.
6. After confirming quality and concentration of the libraries, they are submitted for sequencing using Illumina HiSeq 2500 platform or similar.

3.4 SNP Calling and Alignment

1. Prior to alignment to the reference genome, low quality reads should be removed. Trimming can be performed using Trimmomatic v0.32; however, similar trimming programs could be used.
2. Reads containing Ns need to be removed, and then adaptor sequences trimmed.
3. Low quality nucleotides of each read (quality value <20) are removed from the 3′ to 5′.
4. Discard reads with less than 36 nt, along with their paired-end reads.
5. Align clean reads to the genome of interest. We used rice reference genome (MSU 7.0) implementing a bowtie2 pipeline (--very-fast-local). For other cereals use the most recent genome version.
6. The generated .bam files then can be submitted to GATK (version 3.5-0-g36282e4) for SNP extraction using the

parameters –baq CALCULATE_AS_NECESSARY -hets 0.0010-stand_call_conf 0 –stand_emit_conf 0.

7. SNPs with coverage above 10 were used for the subsequent analysis.
8. Only uniquely mapped reads need to be used for subsequent analysis. Polymorphisms detected from the alignments were used to develop read counts of the parental alleles (Fig. 2A).

3.5 Statistical Methods for Imprinted Genes Mining

1. To determine potential imprinted genes of plants, a threshold of a two-fold expression than the expected needs to be used. For instance, for a maternally expressed gene, the ratio of the number of maternal transcripts to the number of paternal transcripts needs to be higher than 4:1. Alternatively, for a PEG, the ratio of the number of maternal transcripts to the number of the paternal transcripts needs to be less than 2:2. We recommend a more stringent threshold (>10:1 or <2:5 for the MEGs and PEGs, respectively) to avoid false positive signals.
2. To test whether a gene expression deviation from the 2:1 (maternal:paternal) expectation is significant, *Chi*-square test can be used as a reliable method to confirm imprinted genes with polymorphisms.

3.6 Validation of the Parent-of-Origin-Dependent Expression Pattern

3.6.1 Reverse Transcription PCR (RT-PCR)

1. To validate a gene showing parent-of-origin-dependent expression pattern, design primers spanning a SNP site (Fig. 2B).
2. Amplify parental transcripts from endosperm cDNA of the parental lines and the reciprocal hybrids by RT-PCR approach.
3. Set up RT-qPCR reactions containing: 5 μL of cDNA of endosperm as a template, 10 μL of KAPA SYBR® FAST qPCR Master Mix (2×), 1 μL of each forward and reverse primer at a concentration of 10 μM, and 8 μL of H_2O. Load samples in triplicate in a PCR plate and spin down while avoiding the formation of foam.
4. qPCR cycling conditions are as follows: 3 min at 95 °C, followed by 40 cycles of denaturation at 95 °C for 15 s; annealing at 60 °C for 20 s; and extension at 72 °C for 20 s. Set a melt curve to run at 60 °C and then 10 s each at 0.2 °C increments between 62 and 95 °C.
5. The PCR products amplified from the reciprocal crosses along with those from the parents are sequenced. If only the maternal or paternal alleles are found, as the examples shown in Fig. 2B, it strongly suggests that the genes are MEGs or PEGs, respectively. CAPS assay is an additional technique that might be considered, which allows to distinguish polymorphisms between parental alleles by enzymatic cleavage.

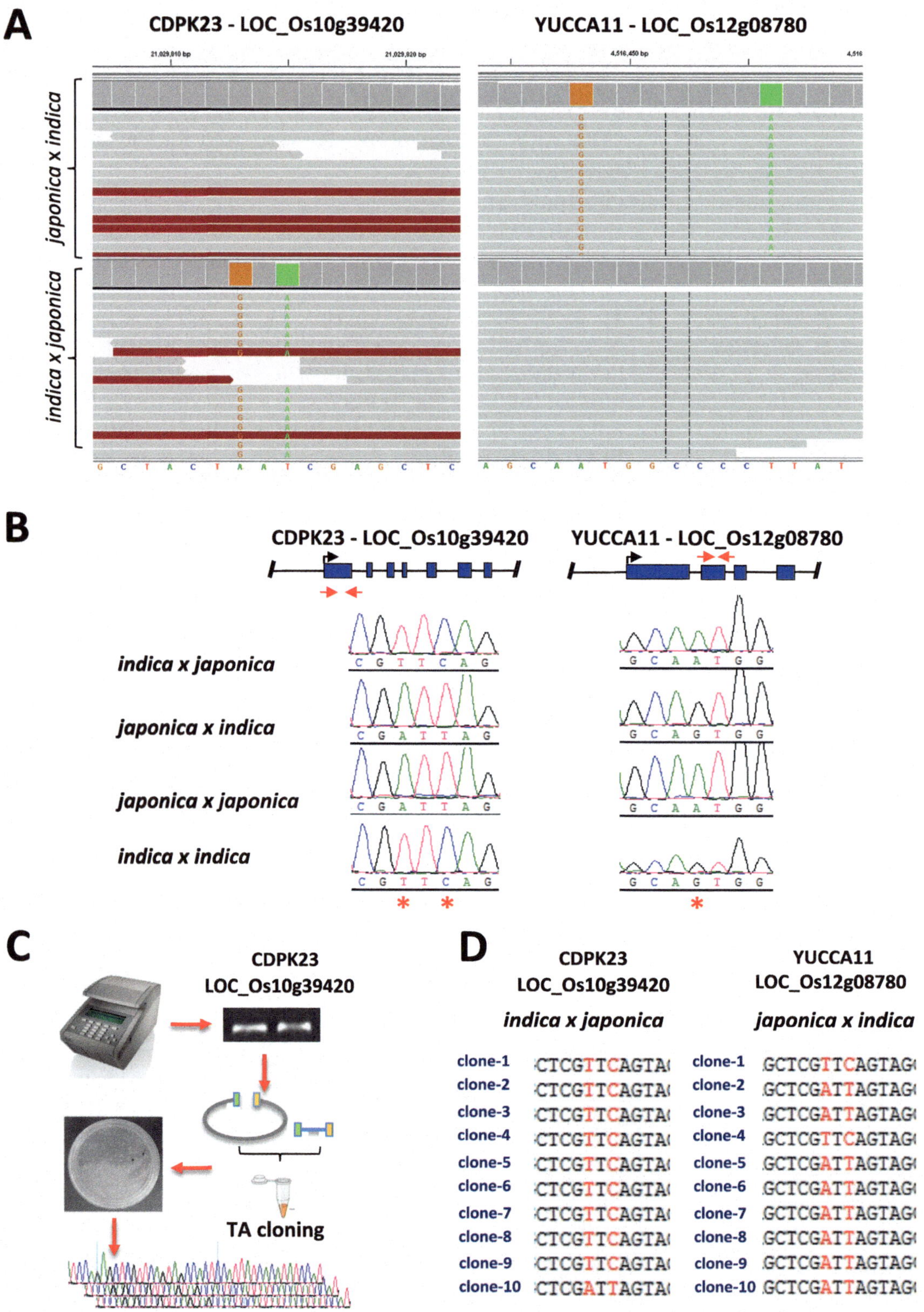

Fig. 2 Identification of imprinted genes in rice. (**A**) *Calcium-dependent protein kinase 23* (*CDPK23*, LOC_Os10g39420) and *Indole-3-pyruvate monooxygenase 11* (*YUCCA11*, LOC_Os12g08780) identified as a MEG and a PEG in rice, respectively. (**B**) Gene model of *CDPK23* and *YUCCA11*; arrows represent primers designed

3.6.2 Cloning and Sequencing

1. Maternal and paternal alleles can also be validated by cloning and sequencing candidate genes. To quantify parental transcripts, qPCR products need to be cloned into a TA vector (Fig. 2C).
2. Ligate qPCR products into pGEM®-T Easy Vector or similar TA cloning vector and the control insert.
3. Set up ligation reactions in 0.5 mL tubes known to have low DNA binding capacity as followed: 5 μL of 2× Rapid Ligation Buffer, 1 μL of pGEM®-T Easy Vector (50 ng), 1 μL of T4 DNA Ligase (3 Weiss units/μL), 2 μL qPCR product and 1 μL of deionized water to a final volume of 10 μL.
4. Mix the reactions by pipetting. Incubate the reactions 1 h at room temperature. Alternatively, incubate the reactions overnight at 4 °C for the maximum number of transformants.
5. Prepare LB/ampicillin/IPTG/X-Gal plates.
6. Spin down the ligation reactions and add 2 μL of each ligation reaction to a sterile 1.5 mL tube on ice. Prepare a control tube with 0.1 ng of uncut plasmid.
7. Use 50 μL of thermocompetent high efficient cells for a better number of transformants. Place competent cells on ice for 5 min.
8. Mix cells by gently pipetting up and down.
9. Carefully transfer each ligation reaction, including controls, to a tube containing 50 μL of competent cells. Mix cells and ligations by gently pipetting up and down.
10. Incubate on ice for 30 min.
11. Heat-shock the competent cells containing the ligation for 45 s in a water bath, thermo block or similar at exactly 42 °C (*see* **Note 12**).
12. Immediately transfer the tubes to ice for 5 min.
13. Add 500 μL LB medium at room temperature to the competent cells containing ligation reactions and incubate them for 1 h at 37 °C on a shaker (~200 rpm).
14. Plate 100 and 200 μL of each transformation culture onto LB/ampicillin/IPTG/X-Gal plates.
15. Incubate plates overnight at 37 °C. Select white colonies. Blue colonies do not contain the insert of interest.

Fig. 2 (continued) over the gene for the SNP identification. SNPs (indicated by stars) located in the coding region were used to distinguish the *japonica* from *indica* parents. For the *japonica* and *indica* reciprocal hybrids (*indica* × *japonica* and *japonica* × *indica*), only maternal or paternal peaks can be observed, indicating the parent-of-origin-dependent expression in the hybrids. (**C**) Overview of one of the validation methods (cloning and sequencing) used for the identification of imprinted genes. (**D**) Clones used to confirm SNPs from maternal and paternal parents

16. Pick single white colonies and grow them in 5 mL of liquid LB medium overnight at 37 °C.
17. Extract plasmid DNA using QIAprep® Miniprep or similar.
18. Send extracted plasmid DNA from each individual clone to DNA sequencing. To evaluate the maternal/paternal ratio use SNPs present in the genome of the parental lines. At least 10 or more individual clones are required to be sequenced per gene to make an unbiased assessment (Fig. 2D).

4 Notes

1. It is necessary to use closely related species with certain level of genomic polymorphisms to be able to distinguish between paternal and maternal genomes.
2. If you use varieties/cultivars with different flowering time, plant them during consecutive weeks to match their flowering windows.
3. Alternatively, 0.5% SDS solution in DEPC-treated water can be used.
4. A cytoplasmic male sterile (CMS) line could be used to prevent auto fertilization. The CMS system contains three types of lines: a CMS line, a maintainer line and a restorer line. A CMS line harbors a sterililty gene encoded in the mitochondrial genome, which leads to complete sterility. A maintainer line does not pose a sterililty gene into its genome; therefore, it is able to produce viable and fertile pollen. Because a CMS lines does not produce viable pollen, when it crosses with a maintainer line, all the seeds obtained are true hybrids. Thus, it avoids false hybrids and greatly reduces labor work.
5. If you use 1.5–2 mL microcentrifuge tube and pestle for homogenization sample the volume should not exceed the 100 μL mark in the Eppendorf tube. First, start with 500 μL of TRIzol, and then add the remaining 500 μL.
6. RNeasy Plant Mini Kit contains a lysis buffer for secondary metabolites-rich tissues. It is recommended for RNA extraction of milky endosperm.
7. Centrifugation separates the mixture into three phases: a red organic phase containing the protein; an interphase, usually white, containing the DNA; and a colorless upper aqueous phase containing RNA.
8. Use a 100–200 μL pipette to avoid contamination picking the interphase containing the DNA during the transferring. Usually, maximum of 700 μL of a colorless phase can be transferred to the clear tube.

9. Do not vortex the samples during this step.
10. A white pellet will form on the bottom or on the side of the Eppendorf tube. If you do not see it, continue to the next step. Sometimes it is transparent and difficult to see.
11. The RNA pellet may float.
12. Do not shake.

Acknowledgments

This work was supported by a departmental start-up grant from the Environmental Horticulture Department at the University of Florida to K.B.

References

1. Rose NR, Klose RJ (2014) Understanding the relationship between DNA methylation and histone lysine methylation. Biochim Biophys Acta 1839:1362–1372
2. Chen C, Li TT, Zhu S et al (2018) Characterization of imprinted genes in rice reveals conservation of regulation and imprinting with other plant species. Plant Physiol 177:1754–1771
3. Raissig MT, Baroux C, Grossniklaus U (2011) Regulation and flexibility of genomic imprinting during seed development. Plant Cell 23:16–26
4. Zhao P, Begcy K, Dresselhaus T et al (2017) Does early embryogenesis in eudicots and monocots involve the same mechanism and molecular players? Plant Physiol 173:130–142
5. Rodrigues JA, Zilberman D (2015) Evolution and function of genomic imprinting in plants. Genes Dev 29:2517–2531
6. Jiang H, Kohler C (2012) Evolution, function, and regulation of genomic imprinting in plant seed development. J Exp Bot 63:4713–4722
7. Waters AJ, Bilinski P, Eichten SR et al (2013) Comprehensive analysis of imprinted genes in maize reveals allelic variation for imprinting and limited conservation with other species. Proc Natl Acad Sci U S A 110:19639–19644
8. Liu J, Li J, Liu HF et al (2018) Genome-wide screening and analysis of imprinted genes in rapeseed (*Brassica napus* L.) endosperm. DNA Res 25:629–640
9. Dong X, Zhang M, Chen J et al (2017) Dynamic and antagonistic allele-specific epigenetic modifications controlling the expression of imprinted genes in maize endosperm. Mol Plant 10:442–455
10. Zhang MS, Li N, He WN et al (2016) Genome-wide screen of genes imprinted in sorghum endosperm, and the roles of allelic differential cytosine methylation. Plant J 85:424–436
11. Gross BL, Zhao Z (2014) Archaeological and genetic insights into the origins of domesticated rice. Proc Natl Acad Sci U S A 111:6190–6197
12. Folsom JJ, Begcy K, Hao X et al (2014) Rice fertilization-Independent Endosperm1 regulates seed size under heat stress by controlling early endosperm development. Plant Physiol 165:238–248
13. Begcy K, Sandhu J, Walia H (2018) Transient heat stress during early seed development primes germination and seedling establishment in rice. Front Plant Sci 9:1768
14. Chen C, Begcy K, Liu K et al (2016) Heat stress yields a unique mads box transcription factor in determining seed size and thermal sensitivity. Plant Physiol 171:606–622

Chapter 12

Analysis of Epigenetic Modifications During Vegetative and Reproductive Development in Cereals Using Chromatin Immunoprecipitation (ChIP)

Kevin Begcy and Thomas Dresselhaus

Abstract

The study of heritable genetic changes that do not implicate alterations in the DNA sequence—epigenetics—represents one of the most prolific and expanding fields in plant biology during the last two decades. With a focus on DNA methylation and histone modifications, recent advances also reported the identification of epigenetic regulatory mechanisms that control reproductive development in cereal crop plants. One of the most powerful methods to selectively study interactions between epigenetic factors or specific proteins bound to genomic DNA regions is called chromatin immunoprecipitation (ChIP). ChIP can be widely used to determine the presence of particular histones with posttranslational modifications at specific genomic regions or whether and where specific DNA-binding proteins including transcription factors interact with candidate target genes. ChIP is also an exciting tool to study and compare chromatin states under normal and stress conditions. Here, we present a detailed step-by-step ChIP assay to investigate epigenetic chromatin marks during vegetative and reproductive development in cereals. However, the method described here can be used for all plant tissues and plant species.

Key words Cereals, Reproduction, Chromatin, Epigenetics, Histone modification, ChIP

1 Introduction

In recent years, epigenetics has promptly developed into one of the most prolific, expanding and influential areas in the biosciences. In plants, a large number of reproductive processes have been also described associated to epigenetic regulation including the transition from diploid sporophytic tissues to the establishment of germ lines and haploid gametophytic generations [1–3]. These include the production of male (pollen) and female (embryo sacs) gametophytes, which upon maturation, are ultimately required in a highly complex and dynamic process of double fertilization that is characteristic of all angiosperms (flowering plants) [4, 5]. Epigenetic processes, including histone methylation, histone acetylation, and DNA methylation, are highly dynamic and have been shown to

Luis M. Vaschetto (ed.), *Cereal Genomics: Methods and Protocols*, Methods in Molecular Biology, vol. 2072,
https://doi.org/10.1007/978-1-4939-9865-4_12,

regulate the switch from vegetative to reproductive development under normal and stressful conditions in model and cereal crop plants [4]. Among the different epigenetic factors, histone modifications including methylation, phosphorylation, acetylation, ubiquitination, and sumoylation are the most frequent epigenetic marks studied. Typically, monomethylated, dimethylated, and trimethylated lysines and arginines are the most widespread histone marks across genomes. For instance, trimethylation of lysine 27 of histone H3 (H3K27me3) is associated with a repressive chromatin state [6, 7], while trimethylation of lysine 4 of histone H3 (H3K4me3) and trimethylation of lysine 36 of histone H3 (H3K36me3) have been linked with an activation of the chromatin [8–10]. To investigate how epigenetic factors control gene expression and regulation, chromatin immunoprecipitation (ChIP) is now the preferred method of choice.

ChIP can be used to study the association of epigenetic factors, modified histones and other DNA-binding proteins at specific DNA regions at a genome-wide scale [11, 12]. ChIP is based on the ability of a crosslinking agent to form stable protein–DNA and protein–protein structures that can be reversed by acid or elevated temperatures. Cross-linking by UV irradiation was initially used [13], but nowadays formaldehyde is the most widely used cross-linking agent [14]. After DNA is ChIPed, it can then be detected by a variety of downstream approaches including semiquantitative PCR, real-time PCR, southern blotting, DNA-Chip or sequencing (Fig. 1). Here we present a detailed protocol for ChIP including PCR and qPCR as downstream detection methods.

2 Materials

2.1 Plant Material

Vegetative and reproductive tissues of cereals such as leaves, roots, whole flowers, glumes, lodicules, anthers, pollen, ovaries, dissected embryo sacs, tassels, seeds, and silks. Several methodologies have been developed for the isolation of a variety of reproductive tissues including male and female gametophytes, embryos and endosperm [15–18].

2.2 Equipment

1. Tissue Lyser II (Qiagen) or a pestle and mortar.
2. Bioruptor® Plus sonication device (Diagenode) or similar.
3. MasterCycler RealPlex (Eppendorf) or similar.
4. Mastercycler ×50 (Eppendorf) or similar.
5. Centrifuge 5424R (Eppendorf) or similar.
6. Growth chambers (I-36VL; Percival Incubators) or similar.

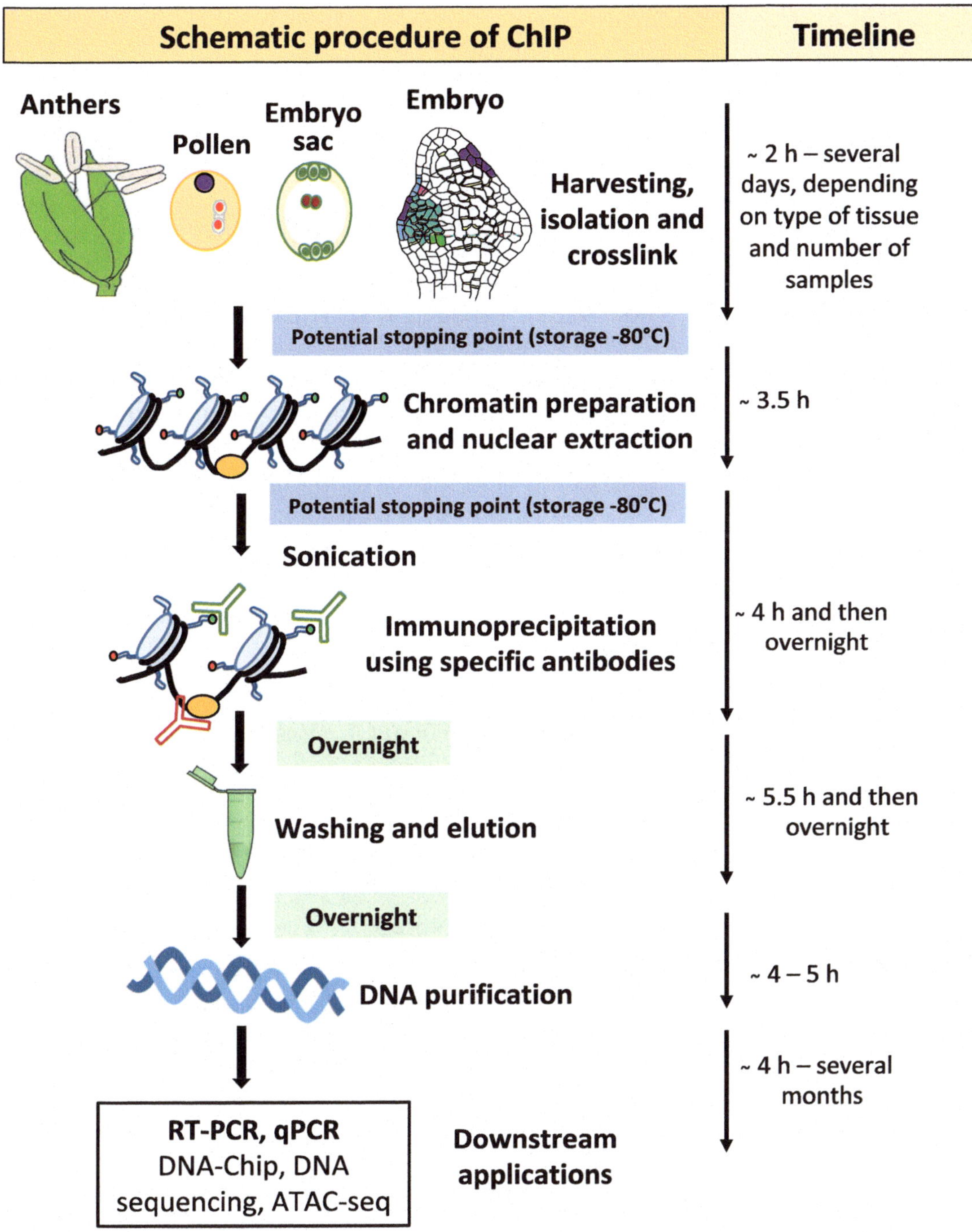

Fig. 1 Schematic illustration and timeline of a chromatin immunoprecipitation (ChIP) protocol designed for plant reproductive development and other tissues

7. Tube Rotator with Free Falcon Tube Holder, 120 VAC (Stuart) or similar.
8. Vacuum pump attached to a desiccator (Cole-Parmer) or similar.
9. Low protein binding collection tube (Thermo Fisher) or similar.
10. Hybridization oven (VWR) or similar.

2.3 Reagents

1. Anti-trimethyl-Histone H3 (Lys27) antibody (Anti-H3K27me3). Rabbit polyclonal antibody. Catalog number: 07-449 (Sigma-Aldrich).
2. Anti-trimethyl-Histone H3 (Lys36) antibody (Anti-H3K36me3). Rabbit monoclonal antibody. Catalog number: 05-801 (Sigma-Aldrich).
3. Anti-Histone H3 antibody (Anti-H3). Rabbit polyclonal antibody. Catalog number: 09-838 (Sigma-Aldrich).
4. Tablets: cOmplete™, EDTA-free Protease Inhibitor Cocktail (Sigma-Aldrich).
5. Phase Lock Gel™, QuantaBio (VWR International).
6. Miracloth (Millipore Sigma).
7. GoTaq® Green Master Mix (Promega).
8. KAPA SYBR® FAST qPCR Master Mix (2×) Universal Kit(Sigma-Aldrich).

2.4 Solutions

Prepare all solutions fresh each time immediately before usage. Keep everything on ice in a cold room. Stocks may be prepared beforehand.

1. Fixation Buffer (*see* **Note 1**)

Solution	Stock	to 37 mL
1% formaldehyde	37%	1 mL
0.4 M sucrose	2 M	6 mL
10 mM HEPES pH 8.0	1 M	0.3 mL
100 mL H_2O + 2 tablets cOmplete™ (EDTA free proteinase inhibitor)	–	30 μL
1× AEBSF (4-(2-aminoethyl) benzenesulfonyl fluoride hydrochloride)	97%	30 μL
5 mM β-ME (mercaptoethanol)	14.3 M	10.5 μL
β-Glycerophosphate	99%	30 μL
H_2O	–	29.59 mL

2. Extraction Buffer 1

Solution	Stock	to 30 mL
0.4 M sucrose	2 M	6 mL
10 mM HEPES pH 8.0	1 M	0.3 mL
100 mL H_2O + 2 tablets cOmplete™	–	30 µL
1× AEBSF	97%	30 µL
5 mM β-ME	14.3 M	10.5 µL
β-Glycerophosphate	99%	30 µL
H_2O	–	23.65 mL

3. Extraction Buffer 2

Solution	Stock	to 1 mL
0.25 M sucrose	2 M	125 µL
1% triton X-100	10%	100 µL
10 mM HEPES pH 8.0	1 M	10 µL
10 mM mgCl_2	1 M	10 µL
1× AEBSF	97%	1 µL
2 mL H_2O + 1 tablet cOmplete™	–	40 µL
5 mM β-ME	14.3 M	0.35 µL
β-Glycerophosphate	99%	10 µL
H_2O	–	703.65 µL

4. Extraction Buffer 3

Solution	Stock	to 1 mL
1.7 M sucrose	2 M	850 µL
10 mM HEPES pH 8.0	1 M	10 µL
5 mM β-ME	14.3 M	0.35 µL
1% triton X-100	10%	15 µL
2 mM mgCl2	1 M	2 µL
1× AEBSF	97%	1 µL
2 ml H_2O + 1 tablet cOmplete™	–	40 µL
β-Glycerophosphate	99%	10 µL
H_2O	–	71.65 µL

5. Nuclei Lysis Buffer

Solution	Stock	to 1 mL
10 mM EDTA	0.5 M	20 μL
50 mM HEPES pH 8.0	1 M	50 μL
1% SDS	10%	100 μL
1× AEBSF	97%	1 μL
2 mL H_2O + 1 tablet cOmplete™	–	40 μL
β-Glycerophosphate	99%	10 μL
H_2O	–	779 μL

6. ChIP Dilution Buffer

Solution	Stock	to 5 mL
10 mM EDTA	0.5 M	12 μL
16.7 mM HEPES pH 8.0	1 M	83.5 μL
1% triton X-100	10%	550 μL
167 mM NaCl	5 M	167 μL
1× AEBSF	97%	5 μL
2 mL H_2O + 1 tablet cOmplete™	–	40 μL
β-Glycerophosphate	99%	10 μL
H_2O	–	4.18 mL

7. Low Salt Buffer

Solution	Stock	to 5 mL
0.1% SDS	10%	50 μL
1% triton X-100	10%	500 μL
150 mM NaCl	5 M	150 μL
2 mM EDTA	0.5 M	20 μL
20 mM HEPES pH 8.0	1 M	100 μL
H_2O	–	4.18 mL

8. High Salt Buffer

Solution	Stock	to 5 mL
0.1% SDS	10%	50 μL
1% triton X-100	10%	500 μL

Solution	Stock	to 5 mL
2 mM EDTA	0.5 M	20 μL
20 mM HEPES pH 8.0	1 M	100 μL
500 mM NaCl	5 M	500 μL
H_2O	–	3.83 mL

9. LiCl Buffer

Solution	Stock	to 5 mL
1 mM EDTA	0.5 M	10 μL
1% NP40	10%	500 μL
1% sodium deoxycholate	–	50 mg
10 mM HEPES pH 8.0	1 M	50 μL
500 mM NaCl	8 M	156,25 μL
H_2O	–	4.28 mL

10. TE Buffer

Solution	Stock	to 200 mL
1 mM EDTA	0.5 M	0.4 mL
10 mM Tris–HCl pH 8	1 M	2 mL
H_2O	–	197.6 mL

11. Elution Buffer

Solution	Stock	to 10 mL
0.1 mM $NaHCO_3$	1 M	1 mL
1% SDS	10%	1 mL
H_2O	–	8 mL

12. Protein Degradation Buffer

Solution	Stock	to 31 μL
0.5 M EDTA	0.5 M	10 μL
1 M Tris–HCl pH 6.5–6.8	1 M	20 μL
20 mg/mL proteinase K	20 mg/mL	1 μL

3 Methods

Germinate cereals seeds in their specific growth conditions containing a substrate and soil mixture. For rice (*Oryza sativa*), geminate seeds on filter paper moistened with autoclaved double-distilled water (ddH_2O) under dark conditions in growth chambers at a constant temperature of 28 °C [19]. Consider a seed germinated when the coleoptile has clearly emerged. Transplant germinated seeds to soil and grow them under controlled conditions (27–30 °C) in the greenhouse [7, 20]. For wheat (*Triticum aestivum* L.), germinate seeds on filter paper under dark conditions at room temperature in an incubator for 3 to 5 days [21]. Check periodically that the filter paper has not dried out (*see* **Note 2**). For winter varieties, vernalize seedlings at 4 °C for 6 weeks and then transplant them to a one gallon pot of soil-sand mixture (3:1, v/v) and grow them in a growth chamber under the following conditions: relative humidity of 50–70%; 16-h light/8-h dark photoperiod; 21 °C/18 °C day/night temperature [22]. For maize (*Zea mays*), soak seeds in tap water for 12 h and then transfer them to petri dishes. The protrusion of the radicle occurs between 48 and 72 h after seeds being placed on the petri dishes. Transfer germinated seeds in an incubator and then place individual seedling to pots (10 cm diameter, 10 seedlings per pot) containing a substrate and soil mixture (1:1, v/v) to the greenhouse under controlled conditions of 14 h of light at 25 ± 2 °C and 10 h of darkness at 21 ± 2 °C and a constant air humidity of 60–65% [15].

After providing optimal plant growth conditions, allow plants to reach vegetative or reproductive development at the specific stage of interest.

To achieve good quality results, all reactions and incubation steps are performed at room temperature if not stated otherwise explicitly.

3.1 Chromatin Crosslinking

1. Harvest 1–3 g of plant material (anthers, pollen, ovules, silks, seeds, leaves, roots, etc.) into 50 mL Falcon tubes containing prechilled 37 mL of Fixation buffer (*see* **Note 3**).
2. Gently submerge plant material to the bottom of 50 mL Falcon tubes and vacuum infiltrate material for 20 min at 25 pound-force per square inch (PSI).
3. Stop crosslinking by adding 2.5 mL of 2.5 M glycine to 50 mL Falcon tubes containing plant material and vacuum infiltrate samples for 15 min at 25 PSI.
4. Remove material from the vacuum and rinse off formaldehyde (**Fixation Buffer**) with 40 mL Milli-Q water. Repeat three times. After the last wash, remove as much water as possible (*see* **Note 4**).

5. Freeze plant material in liquid nitrogen. At this step material can be stored at −80 °C until further use.

3.2 Chromatin Preparation

1. Transfer plant material to a Tissue Lyser II adapter prechilled in liquid nitrogen and homogenize with a frequency of 30 Hz for 1 min (*see* **Note 5**). Alternatively, plant material can be ground using a pestle and mortar to a fine powder.
2. Add ground material to 30 mL of Extraction Buffer 1 in a 50 mL Falcon tube and incubate on a rotating wheel at 4 °C for 10 min. Do not vortex.
3. Centrifuge the solution for 20 min at 3000 *g* at 4 °C.
4. Filter the solution through a double layer of Miracloth into a clean 50 mL Falcon tube (*see* **Note 6**).
5. Centrifuge the filtered solution for 20 min at 3000 *g* at 4 °C.
6. Gently remove the supernatant and resuspend the pellet containing crude nuclei in 1 mL of Extraction Buffer 2 by pipetting up and down.
7. Transfer the solution to a 2 mL Eppendorf tube and centrifuge at 12,000 *g* for 10 min at 4 °C. Repeat this step, 3 to 4 times (*see* **Note 7**).
8. Remove supernatant and resuspend the pellet in 400 μL of Extraction Buffer 3 (*see* **Note 8**).
9. In a clean 2 mL Eppendorf tube add 1.1 mL of Extraction Buffer 3 and gently layer the resuspended pellet on top (*see* **Note 9**).
10. Centrifuge for 1 h at maximum speed at 4 °C.
11. Remove the supernatant and resuspend the chromatin pellet in 400 mL of Nuclei Lysis Buffer. Transfer resuspended pellet to a clean 2 mL Eppendorf tube. Incubate on ice for at least 30 min.

3.3 Sonication

1. Collect 50 μL (before sonication) to compare with the sonicated material.
2. Sonicate samples for 10 cycles, program: 30 s on/off (*see* **Note 10**).
3. Collect 50 μL (after sonication) to compare sonication efficiency (*see* **Note 11**).
4. Centrifuge remaining chromatin solution for 10 min at 14,000 *g* at 4 °C. Transfer supernatant to a clean 1.5 mL low protein binding collection tube. Repeat centrifugation once with aforementioned parameters.
5. In a low protein binding tube wash Protein A agarose beads three times in ChIP Dilution buffer. After each washing step, collect pellet beads by centrifugation for 1 min at 2000 *g* and discard the supernatant.

6. Incubate the chromatin solution from **step 4 (from section 3.3 Sonication)** together with the washed protein A agarose beads from **step 5 (from section 3.3 Sonication)** for 2 h at 4 °C.
7. Centrifuge for 1 min at 2000 *g* and transfer the supernatant into a clean 15 mL tube.

3.4 Immunoprecipitation (IP)

1. Dilute the supernatant 10× with ChIP Dilution Buffer.
2. In several low protein binding tubes divide the diluted chromatin solution. Use 300–500 μL per immunoprecipitation (IP) reaction.
3. Add desired antibodies to each tube, with the exception of the mock control (no antibody). Leave the tubes rotating overnight at 4 °C (*see* **Note 12**).

3.5 Washing

1. In a low protein binding tube wash 40 μL protein A agarose beads three times with ChIP Dilution Buffer; pellet beads by centrifuging for 1 min at 2000 *g* at 4 °C. Prepare a tube for each antibody used and the mock control.
2. Add overnight-incubated chromatin solution containing antibodies to the pelleted beads at 4 °C for 3 h with gentle rotation.
3. Centrifuge for 1 min at 2000 *g* at 4 °C.
4. Collect 500 μL of the supernatant each of the mock control samples as input DNA.
5. Carefully discard remaining supernatants.
6. Wash pellet beads with 500 μL of Low Salt Buffer for 10 min at 4 °C with gentle rotation.
7. Centrifuge for 1 min at 2000 *g* at 4 °C.
8. Wash pellet beads with 500 μL of High Salt Buffer for 10 min at 4 °C with gentle rotation.
9. Centrifuge for 1 min at 2000 *g* at 4 °C.
10. Wash pellet beads with 500 μL of LiCl buffer for 10 min at 4 °C with gentle rotation.
11. Centrifuge for 1 min at 2000 *g* at 4 °C.
12. Wash pellet beads twice with 500 μL of TE buffer for 10 min at room temperature with gentle rotation.
13. Centrifuge for 1 min at 2000 *g* at 4 °C.

3.6 Elution

1. Remove TE buffer and elute immune complexes by adding 250 μL of freshly made Elution Buffer to beads.
2. Vortex briefly and incubate for 15 min at 65 °C with gentle rotation.

3. Pellet beads by centrifugation for 1 min at 2000 *g* at 4 °C.
4. Carefully transfer the supernatant to another tube. Repeat elution step (**steps 1, 2** and **3** from section 3.6 Elution) and combine tubes (500 μL total).
5. Reverse crosslinking of all the samples (*see* **Note 13**). Add 20 μL of 5 M NaCl to the eluate and incubate overnight at 65 °C.

3.7 DNA Extraction and Signal Quantification

1. Add 31 μL of freshly prepared Protein Degradation Buffer to the eluate. Incubate for 3 h at 45 °C with agitation in a hybridization oven.
2. Recover DNA using phase lock gel tubes. Spin down phase lock gel tubes for 30 s at 16,000 *g*.
3. Add 500 μL of 25:24:1 Phenol/Chloroform/Isoamyl alcohol and shake tubes for 10 s.
4. To separate phases centrifuge for 5 min at 16,000 *g*.
5. Add 500 μL of 1-Brom-3-Chloropropan and centrifuge for 5 min at 16,000 *g*.
6. Take the top 400 μL of each tube containing ChIPed DNA and transfer to a clean tube.
7. Add 40 μL of 3 M NaAc pH 5.8 and then 1 mL of 100% ethanol.
8. Place samples at −80 °C for 1 h.
9. Centrifuge at maximum speed for 30 min at 4 °C.
10. Wash pellet with 1 mL of 70% ethanol. Centrifuge for 5 min. Discard supernatant and let ethanol dry at room temperature for 5 min.
11. Add 40 μL of H_2O or TE elution buffer with 10 μg/mL of RNase to dissolve DNA (*see* **Note 14**).
12. Extracted ChIP samples can be stored at −20 °C until ChIP semiquantitative PCR (ChIP-PCR), ChIP quantitative real-time PCR (ChIP-qPCR), or sequencing analysis are performed.
13. Prior to any downstream analysis ChIPed DNA should be diluted. Typically, a 1:10 dilution of the eluate is sufficient, however a larger dilution might be necessary when higher concentration of DNA is obtained.

3.8 Signal Quantification

3.8.1 ChIP Semiquantitative PCR (ChIP-PCR)

1. For ChIP-PCR, specific primers should be designed to encompass the DNA region of interest. Since every gene likely contains variable levels of each epigenetic mark through their promoter and gene body in every cell, it is necessary to test different regions in order to ensure proper amplification of fragmented ChIPed DNA (Fig. 2a).

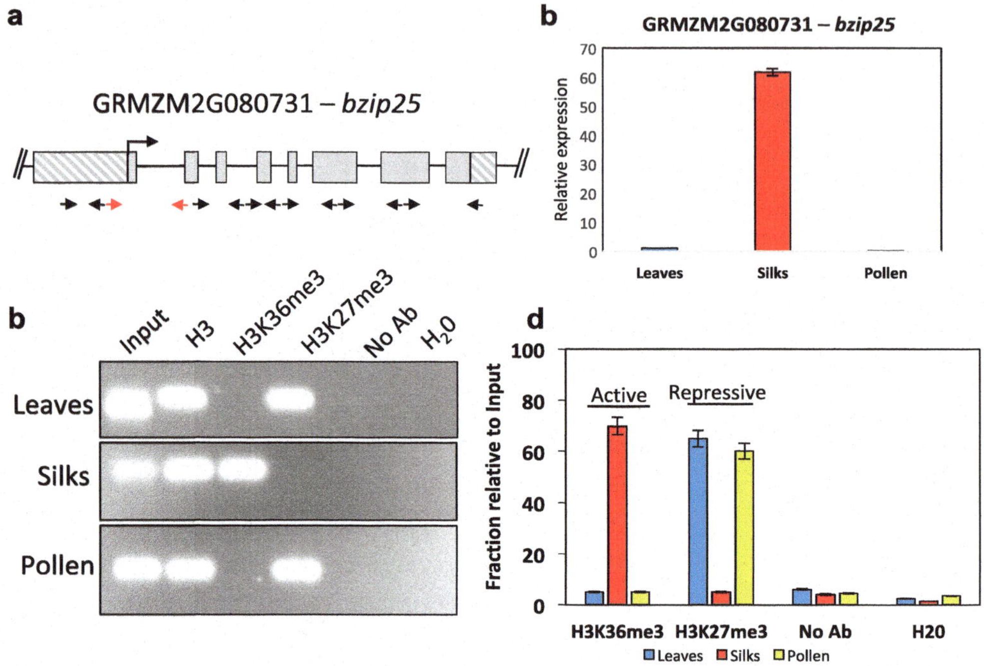

Fig. 2 Chromatin immunoprecipitation (ChIP) analysis of a maize silk-specific gene *bzip25* (GRMZM2G080731). (**a**) Gene model of *bzip25*; arrows represent primers designed over the gene. Red arrows represent set of primers used for ChIP-PCR and ChIP-qPCR shown in (**b–d**). (**b**) *bzip25* gene expression in leaves, silks, and pollen. (**c**) ChIP-PCR analysis of *bzip25* in leaves, silks, and pollen. H3 was used as an additional control. (**d**) ChIP-qPCR analysis of *bzip25* of the active epigenetic chromatin mark H3K36me3 and the repressive chromatin mark H3K27me3

2. Small amount of DNA material is needed for ChIP semiquantitative PCR. Dilute ChIP samples by adding 6 μL of ChIP DNA to 94 μL of H_2O or TE elution buffer. Use 5 μL of diluted sample per PCR reaction.
3. For each reaction, mix 12.5 μL of PCR GoTaq® green Master mix with 1 μL of 10 μM specific forward and reverse primers in a total volume of 25 μL. Alternatively, 1 U Taq DNA polymerase in combination with commercial reaction buffer mixed with 0.85 μL of 10 mM dNTPs together with primers can be used.
4. Load your semiquantitative PCR reaction in a PCR tube, and run a PCR program in a suitable thermocycler as follows: a first cycle of 10 min at 95 °C, 30 s at 65 °C, and 30 min at 72 °C followed by 30 s at 95 °C, 30 s at 65 °C, and 1 min at 30 °C for 30 cycles (*see* **Note 15**).
5. Run all reactions in 1.5% agarose gel (Fig. 2b).

3.8.2 ChIP Quantitative Real-time PCR (ChIP-qPCR)

1. A more precise alternative is the use of quantitative real-time PCR (qPCR) (Fig. 2c). Similarly, as described in **step 1** (from section 3.8.1 ChIP Semiquantitative PCR (ChIP-PCR)), specific primers need to be designed that include the DNA region of interest (*see* **Note 16**).
2. Set up RT-qPCR reactions containing: 5 μL KAPA SYBR® FAST qPCR Master Mix (2×) Universal Kit, 0.5 μL of each forward and reverse primer at a concentration of 10 μM, 1 μL H_2O and 3 μL of ChIPed DNA. Load samples in triplicate in a PCR plate and spin down while avoiding the formation of foam.
3. qPCR cycling conditions are as follows: 3 min at 95 °C, followed by 40 cycles of denaturation at 95 °C for 15 s, annealing at 60 °C for 20 s, and extension at 72 °C for 20 s. Set a final melt curve to run at 60 °C and then 10 s each at 0.2 °C increments between 62 and 95 °C.
4. To analyze ChIP-qPCR data, one of the most common methods used is fold enrichment [23].
5. The fold enrichment represents the amount of ChIPed DNA of each antibody relative to the negative sample (no antibody) or the signal over background.
6. After the ChIP-qPCR has been performed, average the triplicate Ct values for each antibody used as well as the no-antibody control sample.
7. Calculate the difference (delta—Δ) between the Ct average of the antibody of interest and the no-antibody control (mock), as described in the formula:

$$\Delta \,\text{Ct specific antibody} = (\text{Ct IP}) - (\text{Ct no antibody / mock}).$$

8. Calculate the log two fold enrichment ($2^{-(\Delta\,\text{Ct specific antibody})}$) from each antibody relative to the no-antibody/mock.
9. Plot the results of each antibody tested in each tissue used (Fig. 2d).

4 Notes

1. Add always β-mercaptoethanol at the end of the buffer preparation.
2. Some wheat varieties germinate better if placed at 4 °C rather than room temperature. Germination can be sped up by placing in a germinator (under constant light).
3. Place fresh harvested reproductive plant material into prechilled Fixation Buffer.
4. Rinse plant material properly. Gently dry on a tissue paper until the majority of water is removed. Tissues such as pollen,

embryo sacs and single cell tissues are difficult to dry out using tissue paper, discard all liquid present in the falcon/Eppendorf tubes and let them dry out on air for 3–5 min. Leaves, silks, seeds and any other plant material suitable for harvesting in large quantities should be properly dried with tissue paper.

5. Plant material needs to be homogeneously grounded. An automated device such as a Tissue Lyser II (Qiagen) is recommended ensuring high grinding reproducibility. However, prechilled pestle and mortar could also be used. It is important that all plant material is grounded to a fine powder.
6. This step just filters out any undissolved lumps.
7. To decrease background signal, it is important to resuspend and centrifuge the pellet three to four times until the resulting pellet appears white.
8. If the protocol is working well, this pellet should have a white fleck (nuclei). If it is not white, it may be hidden under other cellular components.
9. Do not mix the two layers. Add the solution on the top carefully. Do not form bubbles in the solution, as this will decrease sonication efficiency.
10. Sonication efficiency highly varies depending on the sonicator device used. Therefore, it is important to determine experimentally sonication conditions ahead. A Bioruptor® Plus sonication device coupled with 1.5 mL tube holder for Bioruptor® Standard & Plus adaptor allows to sonicate multiple samples simultaneously. If sonication is performed using a single sample device, keep samples on ice between sonication to avoid protein denaturation. It is essential to properly sonicate each sample; otherwise long chromatin fragments will decrease ChIP resolution. It is also important to avoid foaming since this will reduce sonication efficiency.
11. In order to test sonication efficiency, incubate 50 μL of supernatant at 65 °C overnight (cross-link reversal) and purify it using the DNA purification steps from section 3.7 DNA Extraction and Signal Quantification. Observe purified non-sonicated and sonicated DNA in a 1.5% agarose gel. Maximum signal intensity is expected at 400 bp of DNA and an optimal average DNA fragment size ranging from 200 to 1000 bp.
12. In order to obtain a high-quality ChIP assay, it is important to experimentally determine the quantity of antibody required. This is especially important for self-made antibodies. Too much antibody can lead to high background noise, while small antibody amounts can cause low or lack of signals. We used 1 μL of commercially available H3 antibody and 3 μL of H3K36me3 as well as H3K27me3 antibodies. Another important aspect to pay attention to is the quality of

the antibodies used. This is a key determinant for a successful ChIP assay. We recommend purification of own, noncommercially available antibodies.

13. Together with the ChIP samples you need to perform reverse crosslinking in all input, before and after sonication samples.
14. Alternative DNA extraction methods may also be used. However, the same method should be used within one experiment.
15. For semiquantitative PCR (ChIP-PCR), sample dilution and number of cycles during PCR reactions have to be adjusted. In an optimal number of cycles, all PCR reactions are still in the exponential phase of amplification. We recommend to perform PCR reactions in a real-time PCR cycler.
16. In order to check the transcriptional expression level of a gene of interest, qPCR could be used as a quantification method.

Acknowledgments

This work was supported by the BayKlimaFit program of the Bavarian State Ministry of the Environment and Consumer Protection (Grant No. 810100) to T.D.

References

1. Zhao P, Begcy K, Dresselhaus T et al (2017) Does early embryogenesis in eudicots and monocots involve the same mechanism and molecular players? Plant Physiol 173:130–142
2. Baroux C, Autran D (2015) Chromatin dynamics during cellular differentiation in the female reproductive lineage of flowering plants. Plant J 83:160–176
3. Russell SD, Jones DS (2015) The male germline of angiosperms: repertoire of an inconspicuous but important cell lineage. Front Plant Sci 6:173
4. Begcy K, Dresselhaus T (2018) Epigenetic responses to abiotic stresses during reproductive development in cereals. Plant Reprod 31:343–355
5. Dresselhaus T, Sprunck S, Wessel GM (2016) Fertilization mechanisms in flowering plants. Curr Biol 26:R125–R139
6. Zhang X, Clarenz O, Cokus S et al (2007) Whole-genome analysis of histone H3 lysine 27 trimethylation in Arabidopsis. PLoS Biol 5:e129
7. Folsom JJ, Begcy K, Hao X et al (2014) Rice fertilization-independent Endosperm1 regulates seed size under heat stress by controlling early endosperm development. Plant Physiol 165:238–248
8. Chen XS, Liu XY, Zhao Y et al (2015) Histone H3K4me3 and H3K27me3 regulatory genes control stable transmission of an epimutation in rice. Sci Rep 5:13251
9. You Q, Yi X, Zhang K et al (2017) Genome-wide comparative analysis of H3K4me3 profiles between diploid and allotetraploid cotton to refine genome annotation. Sci Rep 7:9098
10. Chantalat S, Depaux A, Hery P et al (2011) Histone H3 trimethylation at lysine 36 is associated with constitutive and facultative heterochromatin. Genome Res 21:1426–1437
11. Haque ME, Han B, Wang B et al (2018) Development of an efficient chromatin immunoprecipitation method to investigate protein-DNA interaction in oleaginous castor bean seeds. PLoS One 13:e0197126
12. Milne TA, Zhao K, Hess JL (2009) Chromatin immunoprecipitation (ChIP) for analysis of histone modifications and chromatin-associated proteins. Methods Mol Biol 538:409–423

13. Chodosh LA (2001) UV crosslinking of proteins to nucleic acids. Curr Protoc Mol Biol 36:12.5.1–12.5.8
14. Hoffman EA, Frey BL, Smith LM et al (2015) Formaldehyde crosslinking: a tool for the study of chromatin complexes. J Biol Chem 290:26404–26411
15. Begcy K, Dresselhaus T (2017) Tracking maize pollen development by the leaf collar method. Plant Reprod 30:171–178
16. Gomez JF, Talle B, Wilson ZA (2015) Anther and pollen development: a conserved developmental pathway. J Integr Plant Biol 57:876–891
17. Evans MM (2007) The indeterminate gametophyte1 gene of maize encodes a LOB domain protein required for embryo Sac and leaf development. Plant Cell 19:46–62
18. Li C, Xu H, Russell SD et al (2019) Step-by-step protocols for rice gamete isolation. Plant Reprod 32:5–13
19. Begcy K, Sandhu J, Walia H (2018) Transient heat stress during early seed development primes germination and seedling establishment in rice. Front Plant Sci 9:1768
20. Chen C, Begcy K, Liu K et al (2016) Heat stress yields a unique MADS box transcription factor in determining seed size and thermal sensitivity. Plant Physiol 171:606–622
21. Begcy K, Weigert A, Ogolla Egesa A et al (2018) Compared to Australian cultivars, European summer wheat (Triticum aestivum) overreacts when moderate heat stress is applied at the pollen development stage. Agronomy 8:1–16
22. Begcy K, Walia H (2015) Drought stress delays endosperm development and misregulates genes associated with cytoskeleton organization and grain quality proteins in developing wheat seeds. Plant Sci 240:109–119
23. Haring M, Offermann S, Danker T et al (2007) Chromatin immunoprecipitation: optimization, quantitative analysis and data normalization. Plant Methods 3:11

Chapter 13

Cereal Circular RNAs (circRNAs): An Overview of the Computational Resources for Identification and Analysis

Luis M. Vaschetto, Celso Gaspar Litholdo Jr, Lorena Noelia Sendín, Claudia Mabel Terenti Romero, and María Paula Filippone

Abstract

Circular RNAs (circRNAs) are a widespread class of endogenous noncoding RNAs and they have been studied in the past few years, implying important biological functions in all kingdoms of life. Recently, circRNAs have been identified in many plant species, including cereal crops, showing differential expression during stress response and developmental programs, which suggests their role in these process. In the following years, it is expected that insights into the functional roles of circRNAs can be used by cereal scientists and molecular breeders with the aim to develop new strategies for crop improvement. Here, we briefly outline the current knowledge about circRNAs in plants and we also outline available computational resources for their validation and analysis in cereal species.

Key words Noncoding RNAs, Plant circRNAs, Cereal circRNAs, Computational resources, Molecular breeding, Crop improvement

1 Introduction

Noncoding RNAs (ncRNAs) are important components of the transcriptional regulatory machinery. First identified in human cells [1, 2], circular RNA (circRNA) molecules represent a widespread class of endogenous ncRNAs, being identified in different species ranging from humans and animals to plants [3–6]. They are closely associated to important regulatory roles in animal development [4, 7, 8], and to human diseases, such as Alzheimer and cancer [9, 10]. Although these molecules (circRNAs) have been reported to be abundantly expressed in distinct plant tissues and display evolutionary conservation, their roles in regulation of plant development and stress responses remain largely unknown. In cereals such as rice (*Oryza sativa* L.) [11, 12], maize (*Zea mays* L.) [13, 14], barley (*Hordeum vulgare* L.) [15], and wheat

Luis M. Vaschetto (ed.), *Cereal Genomics: Methods and Protocols*, Methods in Molecular Biology, vol. 2072, https://doi.org/10.1007/978-1-4939-9865-4_13, © Springer Science+Business Media, LLC, part of Springer Nature 2020

(*Triticum aestivum* L.) [16], circRNAs have been identified and experimentally verified. Here, we bring together some of the latest insights into plant circRNAs and outline the available in silico tools for the identification and functional characterization of these molecules in cereal plant species.

2 Classification and Biogenesis

Based on their genomic origin, circRNAs are classified into exonic, intronic, exon-intronic, and intergenic. CircRNAs are generated by different mechanisms, but the most widely studied is referred to as "backsplicing," in which a downstream 5′ splice site (donor) is covalently joined with an upstream 3′ splice site (acceptor) forming a closed loop structure, thereby resulting in a circular RNA molecule (for more detail about the biogenesis of circRNAs in plants *see* [5] and [11]). Figure 1 shows how exonic circRNAs are generated through backsplicing and the comparison of this mechanism with the canonical mRNA splicing pathway. The circRNAs form closed loops in which lack 5′-3′ ends and poly(A) tails, therefore being resistant to exonucleolytic degradation [17, 18]. In animals, both cis-regulatory elements and trans-acting factors, capable of regulating circRNA biogenesis, have been identified [19]; however, components known to mediate biogenesis mechanism in plants have not yet been characterized. It has been shown that circRNAs from animals may be associated with repeats sequences and transposons, which is a requirement for circRNA production [20, 21]; however, in plants it suggests relatively less association with repetitive sequences when compared to animals [5]. Nevertheless, it has been recently suggested that LINE1-like elements and their reverse complementary sequences might affect the formation of circRNAs in maize [13]. Moreover, the biogenesis of circRNAs can be different among plant species, regulated by different factors, and it has been proposed that this mechanism is more complex in maize than in Arabidopsis [22].

3 Biological Functions of Cereal circRNAs

Recent studies suggest that circRNAs may have different roles in the plant development and stress responses [11, 16]. Table 1 indicates studies in cereals that link the expression of circRNAs to distinct biological functions. It has been shown that circRNAs can play important regulatory roles by targeting microRNAs (miRNAs), to act as "miRNA sponges," or by acting as competing endogenous RNAs (ceRNAs) and controlling alternative splicing. [8, 23–25]. In wheat, six dehydration responsive cir-

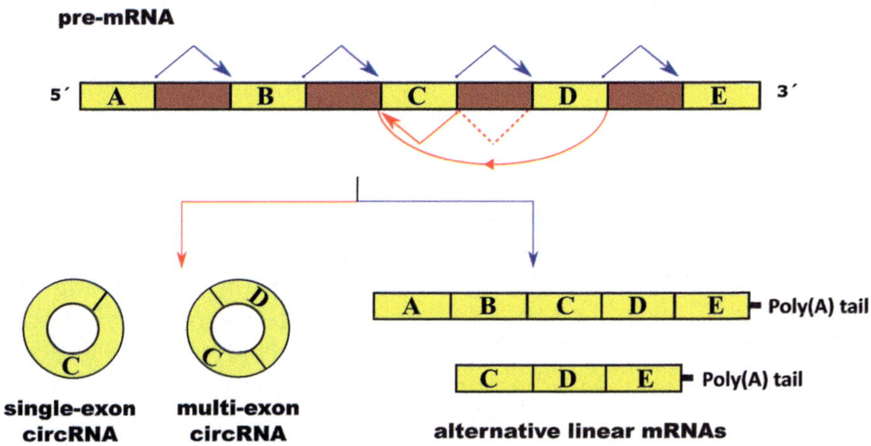

Fig. 1 Exonic circular RNA (circRNA) generated by backsplicing. Exonic circRNAs can be produced by backsplicing of exons of protein-coding genes. Conversely, in the canonical mRNA splicing pathway, linear mRNAs are generated by alternative splicing of consecutive exons. Arrows in the pre-mRNA indicate 5′/3′ splice sites. The red diagonal dashed line indicates the splicing event between exons C and D of the multiexon circRNA. References: exons (yellow); introns (brown)

Table 1
Biological functions associated with the differential expression of circRNAs in cereals

Species	Function	References
Oryza sativa	Regulation of genes involved in metabolic processes Abiotic stress (phosphate imbalance)	Lu et al. [11] Ye et al. [5]
Hordeum vulgare	Micronutrients; mitochondrial function	Darbani et al. [15]
Triticum aestivum	Abiotic stress (dehydration) Root length	Wang et al. [16] Xu et al. [27]
Zea mays	Virus infection Ear height Abiotic stress (drought, heat, and cold)	Ghorbani et al. [14] Chen et al. [13] Tang et al. [26]

cRNAs were reported as miRNA sponges, sequestering the complementary miRNAs, and thus transcriptionally regulating downstream pathways of miRNA target genes [16]. In rice, circRNAs that act as negative regulators of their parental genes have been also identified but little enrichment for miRNA target sites were identified [5, 11].

There is an increasing body of evidence suggesting that cereal circRNAs may have important biological functions. In rice, Ye et al. [5] identified 27 differentially expressed exonic circRNAs which might be involved in response to phosphate-stress conditions. In maize, Zhang et al. [22] reported that circRNAs were differentially expressed in drought stressed maize plants. Interestingly, the authors of this work revealed that plant circRNA host genes may be associated with conserved and/or species-specific drought response pathways. Also in maize, 155 circRNAs were found to be upregulated under virus infection, and it has been proposed that these molecules may control the expression of miRNA and parental protein-coding genes [14]. Another recent report suggests that maize circRNAs may even modulate the phenotypic variation [13]. Tang et al. [26] reported 149 maize circRNAs that were differentially expressed in response to abiotic (i.e., heat, cold, and drought) stress conditions. Moreover, in wheat, Wang et al. [16] showed that 62 circRNAs were differentially expressed under drought stress, and Xu et al. [27] identified wheat circRNAs potentially involved in the regulation of the root length. In barley, circRNAs involved in response to iron and zinc micronutrient levels have also been reported [15].

4 In Silico Resources for Identification and Analysis of Plant circRNAs

In this section, we will explore the computational tools and algorithms specifically designed for the identification and functional characterization of plant circRNAs. In the last years, different computational resources and methods have been developed with the aim at identifying and analyzing circRNAs [28]. In plants, it has been shown that circRNAs may be associated with important biological functions such as stress responses, hormonal signaling, photosynthesis, flower development, and even the regulation of mitochondrial genes [11–16, 26]. Herein, we outline some of the most important in silico resources designed to predict circRNAs from plants, and to also further analyze their putative roles in cellular signaling pathways.

4.1 PlantcircBase 3.0

PlantcircBase 3.0 is a publicly available resource that contains information about plant circRNAs derived from backsplicing and predicted full-length sequences [29]. This resource can be used to predict circRNAs from query sequences suspected to act as miRNA sponges and to regulate the activity of miRNA-mRNA interaction networks. PlantcircBase 3.0 is a user-friendly database that provides useful data on circRNAs identified in different plant species including rice, maize and wheat, being also a multipurpose platform with information of both circRNAs and host genes, additionally including computational tools for the visualization of circRNA structures.

URL: http://ibi.zju.edu.cn/plantcircbase/

4.2 AtCircDB

AtCircDB is a database that contains sequence data of circRNAs obtained by analyzing backsplicing junction sites from *Arabidopsis thaliana*, thereby providing valuable insights for the analysis of circRNAs in other plant species [30]. By using RNA-seq data from different tissues, such as seeds, leaves, and roots, Ye et al. [30] characterized and stored 30648 tissue-specific *A. thaliana* circRNAs and 3486 potential circRNA–microRNA interaction networks—the authors used detection score thresholds in order to categorize the retrieved circRNA sequences.

URL: http://genome.sdau.edu.cn/circRNA

4.3 PcircRNA_finder

The PcircRNA_finder resource was developed with the aim of detecting plant circRNAs through the analysis of RNA-sequencing (RNA-seq) data [31]. It is a platform that optimizes the identification of circRNAs in different plant species [29, 31]. In *A. thaliana* and rice, this tool showed higher sensitivity and precision than find_circ [4] and CIRCexplorer [32], which are two of the most widely used circRNA prediction tools for any organism. PcircRNA_finder is a software especially developed with the aim of identifying exonic circRNAs in plants.

URL (software download): http://ibi.zju.edu.cn/bioinplant/tools/manual.htm

4.4 PlantCircNet

PlantCircNet is a database for the exploration of circRNA–miRNA–mRNA networks and the evaluation of miRNA target sites in crop species, including rice, wheat and maize [33]. This database represents a user-friendly tool to query circRNA-associated regulatory gene pathways, and it has already been used to study physiological signaling associated with flowering time, stress responses and photoperiod. PlantCircNet provides, both predicted and experimentally, verified annotations linked to circRNA–miRNA and miRNA–mRNA networks.

URL: http://bis.zju.edu.cn/plantcircnet/index.php

4.5 CropCircDB

CropCircDB is a useful resource that contains information about circRNAs associated with abiotic stresses, in rice and maize. This database exhibits similar features with available circRNA analysis tools [30, 34]. CropCircDB stores experimentally validated circRNA data, identified from large-scale sequencing projects, especially by including information derived from stress-related RNA-seq datasets. It is expected that CropCircDB may be expanded in the near future to include other crops besides maize and rice.

URL: http://deepbiology.cn/crop/

5 Concluding Remarks

NcRNAs are important transcriptional regulators capable of modulating gene expression at both transcriptional and posttranscriptional levels [35–38]. In particular, circRNAs have recently emerged as a novel class of regulatory ncRNAs which plays important roles in a wide range of biological functions [39]; however, there is still much to learn about their functional nature, especially in plants, where these molecules have been experimentally identified to show differential expression profiles under distinct developmental and stress signaling conditions [16, 40]. The computational tools outlined here have proved to be very useful not only in identifying but also in analyzing the biological functions of plant circRNAs. Further research is necessary and will enable us to better understand the biological relevance of circRNAs, by controlling gene expression, and subsequently apply this knowledge in crop improvement programs.

References

1. Hsu MT, Coca-Prados M (1979) Electron microscopic evidence for the circular form of RNA in the cytoplasm of eukaryotic cells. Nature 280:339–340
2. Cocquerelle C, Daubersies P, Majérus MA et al (1992) Splicing with inverted order of exons occurs proximal to large introns. EMBO J 11:1095–1098
3. Salzman J, Gawad C, Wang PL et al (2012) Circular RNAs are the predominant transcript isoform from hundreds of human genes in diverse cell types. PLoS One 7:e30733
4. Memczak S, Jens M, Elefsinioti A et al (2013) Circular RNAs are a large class of animal RNAs with regulatory potency. Nature 495:333–338
5. Ye CY, Chen L, Liu C et al (2015) Widespread noncoding circular RNAs in plants. New Phytol 208:88–95
6. Ye CY, Xu H, Shen EH et al (2014) Genome-wide identification of non-coding RNAs interacted with microRNAs in soybean. Front Plant Sci 5:743
7. Hansen TB, Jensen TI, Clausen BH et al (2013) Natural RNA circles function as efficient microRNA sponges. Nature 495:384–388
8. Ashwal-Fluss R, Meyer M, Pamudurti NR et al (2014) circRNA biogenesis competes with pre-mRNA splicing. Mol Cell 56:55–66
9. Lukiw WJ (2013) Circular RNA (circRNA) in Alzheimer's disease (AD). Front Genet 4:307
10. Wang Y, Mo Y, Gong Z et al (2017) Circular RNAs in human cancer. Mol Cancer 16:25
11. Lu T, Cui L, Zhou Y et al (2015) Transcriptome-wide investigation of circular RNAs in rice. RNA 21:2076–2087
12. Ye CY, Zhang X, Chu Q et al (2017) Full-length sequence assembly reveals circular RNAs with diverse non-GT/AG splicing signals in rice. RNA Biol 14:1055–1063
13. Chen L, Zhang P, Fan Y et al (2018) Circular RNAs mediated by transposons are associated with transcriptomic and phenotypic variation in maize. New Phytol 217:1292–1306
14. Ghorbani A, Izadpanah K, Peters JR et al (2018) Detection and profiling of circular RNAs in uninfected and maize Iranian mosaic virus-infected maize. Plant Sci 274:402–409
15. Darbani B, Noeparvar S, Borg S (2016) Identification of circular RNAs from the parental genes involved in multiple aspects of cellular metabolism in barley. Front Plant Sci 7:776
16. Wang Y, Yang M, Wei S et al (2016) Identification of circular RNAs and their targets in leaves of Triticum aestivum L. under dehydration stress. Front Plant Sci 7:2024
17. Zhang Y, Liang W, Zhang P et al (2017) Circular RNAs: emerging cancer biomarkers and targets. J Exp Clin Cancer Res 36:152
18. Sablok G, Zhao H, Sun X (2016) Plant circular RNAs (circRNAs): transcriptional regulation beyond miRNAs in plants. Mol Plant 9:192–194

19. Wang Y, Wang Z (2015) Efficient backsplicing produces translatable circular mRNAs. RNA 21:172–179
20. Jeck WR, Sorrentino JA, Wang K et al (2013) Circular RNAs are abundant, conserved, and associated with ALU repeats. RNA 19:141–157
21. Liang D, Wilusz JE (2014) Short intronic repeat sequences facilitate circular RNA production. Genes Dev 28:2233–2247
22. Zhang P, Fan Y, Sun X et al (2019) A large-scale circular RNA profiling reveals universal molecular mechanisms responsive to drought stress in maize and Arabidopsis. Plant J 98:697. https://doi.org/10.1111/tpj.14267
23. Conn SJ, Pillman KA, Toubia J et al (2015) The RNA binding protein quaking regulates formation of circRNAs. Cell 160:1125–1134
24. Zhou R, Zhu Y, Zhao J et al (2017) Transcriptome-wide identification and characterization of potato circular RNAs in response to *Pectobacterium carotovorum* subspecies brasiliense infection. Int J Mol Sci 19:71
25. Li X, Yang L, Chen LL (2018) The biogenesis, functions, and challenges of circular RNAs. Mol Cell 71:428–442
26. Tang B, Hao Z, Zhu Y et al (2018) Genome-wide identification and functional analysis of circRNAs in Zea mays. PLoS One 13:e0202375
27. Xu Y, Ren Y, Lin T et al (2019) Identification and characterization of CircRNAs involved in the regulation of wheat root length. Biol Res 52:19
28. Jakobi T, Dieterich C (2018) Deep computational circular RNA analytics from RNA-seq data. Methods Mol Biol 1724:9–25
29. Chu Q, Zhang X, Zhu X et al (2017) PlantcircBase: a database for plant circular RNAs. Mol Plant 10:1126–1128
30. Ye J, Wang L, Li S et al (2017) AtCircDB: a tissue-specific database for Arabidopsis circular RNAs. Brief Bioinform 20:58. https://doi.org/10.1093/bib/bbx089
31. Chen L, Yu Y, Zhang X et al (2016) PcircRNA_finder: a software for circRNA prediction in plants. Bioinformatics 32:3528–3529
32. Zhang XO, Dong R, Zhang Y et al (2016) Diverse alternative back-splicing and alternative splicing landscape of circular RNAs. Genome Res 26:1277–1287
33. Zhang P, Meng X, Chen H, et al (2017) PlantCircNet: a database for plant circRNA–miRNA–mRNA regulatory networks. Database. 2017. https://doi.org/10.1093/database/bax089
34. Liu YC, Li JR, Sun CH et al (2016) CircNet: a database of circular RNAs derived from transcriptome sequencing data. Nucleic Acids Res 44(D1):D209–D215
35. Amaral PP, Dinger ME, Mattick JS (2013) Non-coding RNAs in homeostasis, disease and stress responses: an evolutionary perspective. Brief Funct Genomics 12:254–278
36. Celton JM, Gaillard S, Bruneau M et al (2014) Widespread anti-sense transcription in apple is correlated with siRNA production and indicates a large potential for transcriptional and/or post-transcriptional control. New Phytol 203:287–299
37. Weiberg A, Jin H (2015) Small RNAs—the secret agents in the plant–pathogen interactions. Curr Opin Plant Biol 26:87–94
38. Zhu B, Yang Y, Li R et al (2015) RNA sequencing and functional analysis implicate the regulatory role of long non-coding RNAs in tomato fruit ripening. J Exp Bot 66:4483–4495
39. Lai X, Bazin J, Webb S et al (2018) CircRNAs in plants. Adv Exp Med Biol 1087:329–343
40. Li QF, Zhang YC, Chen YQ et al (2017) Circular RNAs roll into the regulatory network of plants. Biochem Biophys Res Commun 488:382–386

Chapter 14

Emerging Genome Engineering Tools in Crop Research and Breeding

Andriy Bilichak, Daniel Gaudet, and John Laurie

Abstract

Recent advances in genome engineering are revolutionizing crop research and plant breeding. The ability to make specific modifications to a plant's genetic material creates opportunities for rapid development of elite cultivars with desired traits. The plant genome can be altered in several ways, including targeted introduction of nucleotide changes, deleting DNA segments, introducing exogenous DNA fragments and epigenetic modifications. Targeted changes are mediated by sequence specific nucleases (SSNs), such as zinc-finger nucleases (ZFNs), transcription activator-like effector nucleases (TALENs), and CRISPR (clustered regularly interspersed short palindromic repeats)-Cas (CRISPR associated protein) systems. Recent advances in engineering chimeric Cas nucleases fused to base editing enzymes permit for even greater precision in base editing and control over gene expression. In addition to gene editing technologies, improvement in delivery systems of exogenous DNA into plant cells have increased the rate of successful gene editing events. Regeneration of fertile plants containing the desired edits remains challenging; however, manipulation of embryogenesis-related genes such as BABY BOOM (BBM) has been shown to facilitate regeneration through tissue culture, often a major hurdle in recalcitrant cultivars. Epigenome reprogramming for improved crop performance is another possibility for future breeders, with recent studies on *MutS HOMOLOG 1* (*MSH1*) demonstrating epigenetic-dependent hybrid vigor in several crops. While these technologies offer plant breeders new tools in creating high yielding, better adapted crop varieties, constantly evolving government policy regarding the cultivation of plants containing transgenes may impede the widespread adoption of some of these techniques. This chapter summarizes advances in genome editing tools and discusses the future of these techniques for crop improvement.

Key words Genome editing, Sequence specific nucleases, CRISPR/Cas9, BABY BOOM, Epigenome editing

1 Introduction

Since domestication, conventional breeding has significantly improved crop yields, tolerance to both biotic and abiotic stresses and overall hardiness. This method relies on available genetic and phenotypic diversity from within breeding populations and wild relatives where agronomically important traits are incorporated into elite cultivars. Major drawbacks of the method are the limited

Luis M. Vaschetto (ed.), *Cereal Genomics: Methods and Protocols*, Methods in Molecular Biology, vol. 2072,
https://doi.org/10.1007/978-1-4939-9865-4_14,

number of valuable alleles in breeding populations, difficulty in introducing traits selectively from landraces into elite cultivars and the lengthy process of generating homozygous lines ready for commercialization.

Genetic engineering is an umbrella term that covers precise modifications to genomes by means of targeted DNA insertion, DNA replacement and DNA editing at select loci. One of the approaches to achieve targeted integration or replacement of an endogenous DNA site is through the homologous recombination (HR) DNA repair pathway. Although achievable in model systems (yeast, chicken DT40 cells, mouse embryonic stem (ES) cells and moss *Physcomitrella patens*), gene replacement through HR has only been done at a very low frequency in vascular plants (10^{-4} to 10^{-6}) [1–4]. A rigorous screening process, like positive/negative selection, must be in place to identify rare gene targeting (GT) events in plant population [5–7]. One of the approaches to increase the level of GT events is through artificial introduction of a double strand break (DSB) at the target site. For instance, overexpression of the yeast rare-cutting endonuclease I-*Sce*-I in plant and animal cells increases the rate of GT events through HR by 1000-folds at the cut locus compared to control [8–11]. If gene knockout or mutation is the desired outcome, simple DNA change at a locus can lead to disruption of the sequence following repair with the error-prone nonhomologous end joining (NHEJ) pathway [12]. To expand on the number of sequences that can be targeted in the genome, sequence specific nucleases (SSNs) have been designed that consist of programmable DNA-binding motifs fused to a non-specific endonuclease domain [13]. The first engineered endonucleases appeared in the late 1990s, and currently there are four types of such nucleases that can be used for genome editing (GE): engineered homing endonucleases/meganucleases (EMNs), Zinc Finger Nucleases (ZFNs), Transcription Activator-Like Effector Nucleases (TALENs) and Clustered Regularly Interspaced Short Palindromic Repeats (CRISPR)/CRISPR-associated 9 (Cas9). All of these have been successfully used in plants to introduce modifications at predefined positions in the genome. Nevertheless, challenges with the design, verification and prohibitive licensing fees associated with some of the engineered endonucleases make their utilization less frequent as compared to other SSNs. In particular, at the present time, TALEN and CRISPR/Cas9 are the most widely used technologies in plants [14]. Successful use of genome editing reagents in plant systems also requires efficient delivery of nucleases into plant cells and eventual regeneration of plants from transformed explants. There is striking variation among different plant species and cultivars in transformation efficiency and their amenability to tissue culture. Recent breakthroughs in the application of early embryogenesis-related genes for improvement of transgenesis and breaking recalcitrance to tissue culture has allowed for utilization of the genes to, potentially, improve efficiency of genome editing in crops [15, 16]. Application of engineered nucleases goes beyond genome editing and can be used for manipulation of gene expression through epigenome editing. Additionally, alteration in the global epigenome profile by using *msh1* mutants has emerged as another promising avenue for crops improvement.

2 Genome Editing Using Sequence-Specific Nucleases

2.1 ZFNs for Trait Improvement in Crops

Zinc finger (ZF) motifs are the most abundant DNA-binding domains present in eukaryotes and therefore were the first chimeric nucleases to be designed [17–19]. Development of SSNs was a gradual process that included decoding interaction patterns of ZF motifs with DNA and developing the most efficient strategy of linking them to nonspecific endonuclease *Fok*I from *Flavobacterium okeanokoites*. The ZF unit consists of either three or four binding modules that can recognize a nucleotide triplet. In turn, two ZF monomers can bind specific 18–24 bp-long sequences which need to be spaced by a 5–6 bp gap for *Fok*I dimerization. ZF nuclease (ZFN) monomers upon landing at the target site create DSB with 4–5 bp 5′-overhangs increasing the chance of introducing mutations when repaired through the NHEJ pathway, as compared to the repair of blunt DNA ends [20].

The first study describing editing of the corn genome using ZFNs was published in 2009 [21]. ZFN-mediated disruption of the *IPK1* gene resulted in the generation of plants producing significant number of seeds with reduced phytate levels, an antinutritional component of feed grain. The frequencies of successful GT events ranged from 3.4 to 100%, depending on the ZFN pair used and the donor template. No off-target mutations were observed at the noncognate homologous sites in T0 plants carrying GT events at the *IPK1* gene. Additionally, successful editing of epigenetic-related genes was described in soybean (*Glycine max*) [22]. The whole-plant transformation of soybean using a cassette under the control of an estrogen-inducible promoter and encoding ZFNs targeting two paralogous genes, *DCL4a* and *DCL4b*, resulted in the recovery of three T0 plants with desired editing events.

Commonly grown elite cultivars usually possess a combination of traits that confer either resistance to pathogens, tolerance to abiotic stressors or high yield. Adding these traits one at a time from either landrace varieties or other cultivars is laborious and time consuming since it requires the involvement of introgression via conventional breeding. To streamline the process of adding new traits/genes to the crop genome, the molecular trait stacking method was established [23]. This can be done by using a transformation vector carrying the trait genes with flanking homology sequences to the target region and a ZFN expression cassette designed to target the desired integration locus. In addition to the transgene and homology arms, the DNA sequence carries modular 'trait landing pads' (TLPs) with ZFN target sites homologous to an incoming DNA. Another co-transformation of transgenic plants with a donor DNA containing the second gene of interest flanked by sequences homologous to the integrated TLP along with the corresponding ZFN expression construct allows for this gene to be precisely integrated at TLP, directly adjacent to the first transgene. This approach allows for precise integration of the transgenes at the chosen location in the genome, where desirable expression of the transgenes is expected without collateral disruption of gene coding elements. More importantly, the transgenes segregate together as a single locus following meiosis, therefore obtaining progeny with the combination of all added traits [23].

2.2 TALENs for Trait Improvement in Crops

Following pioneering studies on ZFNs, genome engineering in plants using TALENs has progressed rapidly [24, 25]. A new class of SSNs was produced by using the TALE DNA-binding motifs from *Xanthomonas* plant pathogens and *Fok*I domain [26–28]. The TALE proteins bind to DNA sequence using a middle region that contains 30 tandem repeats composed of 33–35 amino acids. Each repeat contains mostly consistent amino acid sequence except for two adjacent amino acids (the repeat variable di-residue or RVD) at positions 12 and 13. Distinct RVDs within repeats dictate specificity when binding to the target sequence [26, 29, 30]. The off-target effects of TALENs seem to be fewer than those of ZFNs due to the longer target recognition site [4].

The applicability of TALENs for crop improvement has been clearly shown in several species including soybean, wheat, barley and tomato [31–34]. Soybean double mutants of *fad2-1a* and *fad2-1b* genes were generated that contain low level of polyunsaturated fats, improving economic value with longer oil shelf life and improved oxidative stability. It was also possible to segregate out the TALEN construct during reproduction, thereby allowing for recovery of mutant lines without the introduced transgene.

Genome editing also allows for generation of new traits, not found in nature. A wheat mutant line with heritable broad-spectrum resistance to powdery mildew was developed by simultaneously targeting three *TaMLO* homologous alleles in wheat [32]. In another study, disruption of a *VACUOLAR INVERTASE* gene (*VInv*) that encodes an enzyme involved in hydrolysis of sucrose to glucose and fructose led to increase in the cold storage and processing of potato tubers [35]. Homozygous mutant plants had undetectable levels of reducing sugars that can form a potential carcinogen when reacted with free amino acids upon high-temperature processing. Similar to the soybean study, the authors selected *vinv mutants* without TALEN transgenes. *The edited potato is void of the regulation covering GMO crops in the USA and may soon enter a market as the first crop edited with designed endonucleases.*

2.3 The CRISPR/Cas9 System for Trait Improvement in Crops

The bacterial RNA-guided immune system CRISPR/Cas has been used for genome editing on several crop species [36]. Originally discovered in *E. coli* in 1987, it was not until 2012 when the CRISPR/Cas system was redesigned for genome editing applications [37]. Currently, the most frequently used CRISPR/Cas genome editing tool originates from *Streptococcus pyogenes*, although several alternative Cas enzymes have been described and tested *in planta* from other bacteria [38]. For Cas9 binding, specificity is provided by the so-called "seed sequence" of approximately 12 bases and a short DNA sequence termed a protospacer adjacent motif (PAM). The PAM motif usually contains a sequence of 5′-NGG-3′ (less frequently 5′-NAG-3′ [39]) and is situated downstream of the target DNA [40]. In comparison to ZFN and TALEN technology, CRISPR/Cas relies on simple Watson–Crick base pairing between single guide RNA (sgRNA) and the target DNA sequence, and does not require sophisticated protein engineering steps [41]. The digestion of the target DNA sequence is performed by two cleavage domains (RuvC and HNH) of Cas9. The DSB is produced at a position that is three nucleotides upstream of PAM leaving blunt ends in most of cases [37]. Computational mining of bacterial genomes has uncovered several new programmable class II CRISPR/Cas nucleases, which has recently been classified into subtypes V and VI based on both sequence divergence and function [42]. Whereas DNA targeting nucleases belong to subtype V (Cas12a–e), two RNA targeting nucleases (Cas13a–b) belong to subtype IV. One of the most promising alternatives to Cas9 is Cas12a (formerly Cpf1, CRISPR *Prevotella* and *Francisella* 1) that requires T-rich (-TTTN-) PAM sequences instead of Cas9's G-rich PAM. Unlike Cas9 that generates blunt ends, Cas12a produces DSBs with 4- or 5-nt overhangs, leading to sticky ends [43]. This ability of Cas12a is, apparently, responsible for higher frequencies of targeted insertions and replacements through HR observed in plants [44]. In addition to DNA nuclease activity, Cas12a possesses RNase III activity for unprocessed CRISPR RNA (crRNA) that can be used for multiplex targeting of a number of endogenous sequences using tandemly arrayed pre-crRNA expressing constructs [45]. An ability to simultaneously target multiple loci is especially important for engineering of crops with polyploid genomes [25].

Recent analysis of reports published from 2014 to 2017 related to CRISPR-mediated genome editing of crops revealed that rice (*Oryza sativa*) was the most studied plant followed by tobacco (*Nicotiana* sp.), tomato (*Solanum lycopersicum*), or corn (*Zea mays*) [36, 46]. Out of 52 studies examined, 20 described improvement of yield through GE application. Most of the studies were focused on permanent disruption of endogenous sequences through the NHEJ repair pathway. The CRISPR system has been used for mutation of protein coding sequences [47, 48], promoter regions [49], complete deletion of genes or chromosomal fragments [50], gene insertions or allele replacements [50–52] and in-frame fusion of proteins [52]. Interesting applications include disruption or removal of susceptibility genes (S genes) in crop genomes that reduce host-pathogen interactions causing disease. Examples include mutation of *mildew resistance locus O* (*MLO*) in several plant species that confers partial or complete resistance to powdery mildew [32, 53], disruption of *OsERF922* in rice resulting in *X. oryzae* resistance [48] and mutation of *downy mildew resistance 6* (*DMR6*) and *eukaryotic initiation factor 4E* (*eIF4E*) in tomato and cucumber to increase resistance against bacterial and viral pathogens, respectively [54, 55]. For the most up-to-date review on using CRISPR technology for disease resistance, please refer to [38]. Further application of CRISPR technology has been to mimic domestication in wild relatives. Recently, several groups have reported CRISPR editing in tomato relatives, demonstrating the power of CRISPR to generate important agronomic traits [56–58].

Precise integration of a transgene at a defined location remains challenging. Sophisticated delivery and screening methods must be in place to achieve successful integration of the construct through HR repair mechanisms [59–62]. An alternative method, which benefits from the use of NHEJ repair mechanism, has been proposed that involves targeting of introns adjacent to an exon of interest [51]. Although NHEJ is an error-prone repair mechanism, short modifications in the intron sequence do not affect mRNA processing if the splicing sites remain intact. By supplementing donor DNA it was possible to achieve around 2% of gene replacement and targeting efficiency in rice [51].

One of the biggest concerns regarding the CRISPR/Cas system during its initial development were the relatively high off-target effects reported in animals and more recently in plants [39, 63, 64]. Optimization of the transformation efficiency, expression level and codon usage were proposed as possible solutions to mitigate this problem. High concentrations of Cas9 and sgRNA can cause off-target mutagenesis [39, 65, 66]. Careful selection of the target sites in the genome coupled with highly specific SpCas9 variants with tRNA–sgRNA fusions were proposed as possible solutions against off-target effects [64, 67]. However, this requires the availability of a PAM site at a given sequence, with alternative PAM sites reducing but not fully eliminating the activity of CRISPR/Cas in plants [68]. Overall, in silico examination of nuclear genome sequences from eight representative plant species (*A. thaliana*, *M. truncatula*, *G. max*, *S. lycopersicum*, *B. distachyon*, *O. sativa*, *S. bicolor*, and *Z. mays*) demonstrated an occurrence of the PAM (NGG/NAG) site at a frequency of 5–12 PAMs/100 bp [69]. At such frequencies it is possible to target 85.4–98.9% of the annotated transcriptome with specific sgRNAs. Maize, having the largest genome examined with a high number of homologous genes, had the lowest number of specific sgRNA predicted to target transcription units (only 30%). Therefore similar challenges are expected in wheat and barley sgRNA target prediction given their even larger genomes size [69].

3 Chimeric Cas9-Based Fusion Enzymes for Precise Base Editing and Modulation of Gene Expression

An additional level of precision in altering genomic sequence can be achieved by using base editing enzymes fused to either partial (nickase, nCas9) or fully inactivated (dead, dCas9) Cas9. Recently, two base editors, cytidine and adenine deaminase fusions to Cas9, were developed that proved to be functional in cereals such as rice, wheat and corn [70–72]. Two different variants of cytidine deaminase were tested in plants—one derived from rat cytidine deaminase, APOBEC1 and another from *Petromyzon marinus* [72, 73]. Both studies have clearly demonstrated that a nickase fusion carrying D10A mutation was more efficient in triggering C to T substitutions as compared to dCas9, which has both cleavage domains inactivated. The substitutions were recorded between 3 and 9 nucleotides in the protospacer motif counting from the 5′ region and homozygous rice mutants were obtained in the first generation from *Agrobacterium*-transformed calli [72].

Adenine base editor (ABE) was developed by fusing nCas9 and engineered transfer RNA adenosine deaminase derived from *Escherichia coli*. The resulting chimeric enzyme has an ability to induce A/T to G/C conversion [70, 71]. Several combinations of an adenosine deaminase (ecTadA-ecTadA*) with nCas9 (D10A) fusion were tested and the highest efficiency of base editing in rice protoplasts was achieved with adenosine deaminase domain at the N-terminus of nCas9 followed by three nuclear localization signals (NLSs) at the C-terminus [71]. Comparison of three different sgRNA variant—native, enhanced sgRNA (esgRNA) and tRNA-sgRNA expression system indicated that using esgRNA results in the highest rates of base editing in both rice and wheat protoplasts at the endogenous loci. Furthermore, 20-nucleotide spacer of esgRNA was the most efficient for induction of substitutions at the target site as compared to the shorter versions. It was also possible to obtain homozygous rice mutants for several endogenous loci in the first generation using *Agrobacterium*-mediated calli transformation. In comparison to the cytidine deaminase base editor, the effective deamination window was shorter—from 4 to 8 nucleotides in the protospacer motif. None of the obtained transgenic rice plants contained any indels or undesired edits at the target site.

Modulation of gene expression by using dCas9 chimeric proteins adds another degree of flexibility in applied and functional genomics studies in plants. For instance, activation or downregulation of gene expression can be achieved by bringing either VP64 or SRDX domains at the promoter region of the target gene, respectively [74, 75]. The most recent study on development of transcriptional activators in plants, called CRISPR-Act 2.0, demonstrated that recruiting extra VP64 activator domains to the promoter region resulted in up to 1500-fold activation in transcription of *Fertilization-Independent Seed2* (*FIS2*) gene in Arabidopsis as compared to control [76]. The system relies on using bacteriophage coat protein MS2 together with specific RNA-stem loop aptamer scaffold. MS2 has high affinity to the RNA aptamers that are inserted into sgRNA tetraloop and stem-loop 2 positions to make gRNA 2.0. Modified sgRNA was able to efficiently recruit additional VP64 activator domains that were fused to MS2 protein. CRISPR-Act 2.0 system also proved to be efficient in activation of transcription from endogenous promoters in rice protoplasts, demonstrating its applicability in crops [76].

4 Methods for Delivery of Designed Endonucleases into Plant Cells

Despite success in the application of GE technology for traits improvement in crops, successful delivery of the reagents inside plant cells and efficient regeneration of fertile plants following transformation remains challenging. The SSN can be delivered in the form of DNA, RNA or purified protein [77–79]. The most common methods used for delivery of a transgene encoding for the GE reagents are biolistic transformation, *Agrobacterium*-mediated transformation and polyethylene glycol (PEG) mediated transformation of protoplasts. Although DNA-mediated GE is the preferred method of choice to test the efficiency of SSN *in planta*, such an approach can trigger regulations covering genetically modified (GM) crops in some jurisdictions (e.g., EU), even if the transgene is segregated away through the backcross. Therefore, several reports have addressed this issue by transfection of either mRNA or protein complexes that undergo rapid intercellular degradation following cleavage activity [79–82]. Typically, SSNs are overexpressed and purified from bacterial culture, while sgRNAs are in vitro transcribed and complexed with Cas9 in transfection buffer to form ribonucleoprotein (RNP) complexes. Activity and specificity of the RNPs are first tested in vitro using DNA that carries fragment of endogenous sequence. Eventually, RNP complexes are either loaded on gold particles and delivered using bombardment into explants or transfected into protoplast using PEG [77, 79]. Regardless of transfection method, an established regeneration technique must be in place for obtaining fertile GE plants.

Recently, we have proposed an alternative method for nucleic-acid free genome editing in haploid wheat cells by using cell penetrating peptides (CPPs) [83]. Purified ZFN proteins were complexed with CPPs and delivered into microspores, the predecessors of male gametes that under certain conditions can undergo embryogenesis in vitro to produce green plants (submitted to Sci. Rep.). Indels were recovered at the target sites using NGS. Additionally, we also were able to demonstrate the applicability of genome editing in haploid embryos regenerated from untreated microspores. Working with haploid cells/tissues is desirable for biotechnology applications since it significantly reduces time and cost associated with generation of complete homozygous lines with the desired genetic alterations.

5 Elevating Genome Editing Rate in Crops by Manipulating Expression of Embryogenesis-Related Genes

Successful generation of GE crops remains dependent on efficient tissue culture protocols. Unfortunately, the direct introduction of genetic alterations into elite cultivars is often not feasible due to their recalcitrance to tissue culture. One possible approach to improve plant regeneration via tissue culture is by manipulation expression of embryogenesis-related genes. Examples of such genes include *LEAFY COTYLEDON1* (*LEC1*), *LEAFY COTYLEDON2* (*LEC2*) [84, 85], *PICKLE* (*PKL*) [86], *SOMATIC EMBRYOGENESIS RECEPTOR KINASE* (*SERK*) [87], *AGAMOUS-like 15* (*AGL15*) [88], *FUSCA3* (*FUS3*) [89], *SHOOT MERISTEMLESS* (*STM*) [90], *WUSCHEL* (*WUS*) [91] and *BABY BOOM* (*BBM*) [92]. Among the genes only three—*WUS*, *BBM*, and *LEC*—do not require exogenous application of plant hormones for somatic embryogenesis when overexpressed in transgenic plants. For example, in the inbred corn line PHH5G, a cultivar with poor transformation and regeneration efficiencies, temporary overexpression of *WUS* and *BBM* resulted in the recovery of over 40% of transgenic calli [16]. Eventually, most of the transgenic calli produced green and fertile plants. Furthermore, the authors demonstrated increased transformation and regeneration in sorghum (*Sorghum bicolor*), sugarcane (*Saccharum officinarum*), and rice (*Oryza sativa* ssp. *indica*), with *WUS* and *BBM* overexpression, suggesting a universal applicability for monocot transformation. The *BBM* gene is a master regulator of several embryogenesis-related genes and during constitutive overexpression induces somatic embryo formation on seedlings without exogenous growth regulators or stress treatments [93]. Additionally, transgenic plants suffer from pleiotropic effects and seeds from T1 plants have poor germination rate. Therefore, either an elaborate expression or inducible CRE/lox excision system must be in place for tight regulation of *WUS* and *BBM* genes expression following transformation process [16, 94, 95].

A recent study produced haploid and apomictic seeds in rice following manipulation of *BBM* gene expression in reproductive tissues [96]. Generation of both types of seeds is essential for modern breeding practice. Whereas rapid production of fully homozygous plants can be achieved through induction of doubled haploids, apomixis offers the promise of fixing hybrid vigor in breeding programs. The *BBM1* gene is specifically expressed from the male genome after fertilization and can be an inducer of embryo formation in zygote. Indeed, ectopic expression of *BBM1* gene in the egg cell triggered production of haploid plants at frequency of up to 10% in the first generation and reached around 29% in the second generation of homozygous lines [96]. Furthermore, similar overexpression of *BBM1* gene from the female genome in *MiMe* mutant background, where meiosis is replaced by mitosis (triple knockout of the meiotic genes *REC8*, *PAIR1* and *OSD1*), resulted in production of haploid plants while haploid-induction frequency was maintained for the following two generations. Additionally, it was possible to get diploid plants from doubled haploid parents as a result of asexual reproduction. It would be interesting if similar effects can be achieved in other cereals, such as wheat, by using a *BBM* homolog that our group recently mapped [97].

6 Global and Targeted Epigenome Editing for Crop Improvement

Regulation of gene expression is a complex process involving interactions of transcription factors, chromatin conformational changes and epigenetic marks. The epigenetic regulation of gene expression includes three main components: DNA methylation, histone modifications, and the expression of small RNAs (smRNAs). Factors of the epigenetic regulatory landscape are highly flexible, continuously responding to perturbations in environmental conditions. Practical use of epigenetic reprogramming for improved crop yield and performance comes from studies on *MutS HOMOLOG 1* (*MSH1*) gene, which is unique to plants and encodes a homolog of a bacterial mismatch repair protein [98]. The protein is encoded in the nucleus but is localized to mitochondrial and chloroplast nucleoids and is involved in organelle genome stability [99]. Disruption of *MSH1* gene expression by mutation or RNA interference (RNAi) causes instability in the plastid's genome resulting in several morphological changes in both monocots and dicots [98–100]. Moreover, segregation of the RNAi construct from the sorghum genome was not enough to bring an msh1-induced phenotype back to wild type (WT) and the growth habit persisted for nine consecutive generations [101]. Reciprocal crosses of WT segregants with the dwarf phenotype to WT sorghum resulted in many F_1 progeny that were taller and produced more seeds than WT plants. Surprisingly, the highest yield was obtained in F_3 and F_4 generations distinguishing the phenomenon from heterosis, which is observed in F_1 hybrid plants. No changes in genomic DNA sequence were observed in similar experiment in Arabidopsis, therefore suggesting, alteration in epigenetic regulation as a putative cause of *MSH1*-dependent hybrid vigor effect [102]. Overall, similar studies conducted on tomatoes and soybean suggest that global regulators of the epigenetic landscape in plants have the potential to be used for future breeding purposes [103, 104].

Although reports on targeted modulation of epigenetic marks in plants are scarce, a recent study has shown fusion of human DNA demethylase TEN-ELEVEN TRANSLOCATION1 (TET1cd) to an artificial zinc finger (ZF108) designed to target *FLOWERING WAGENINGEN* (*FWA*) can efficiently promote demethylation at the promoter region causing measurable late-flowering phenotype in Arabidopsis [105, 106]. Surprisingly, the demethylation of the *FWA* promoter was heritable even in the absence of the ZF108-TET1cd transgene, with loss of methylation spanning the entire methylated region of the *FWA* promoter including regions distant from the target site. The demethylation construct was successfully converted into CRISPR/dCas9-based system by using SunTag-TET1cd module [107]. In this approach, dCas9 protein is fused through the C-terminus to several tandem copies of peptide epitopes (GCN4). The epitopes are recognized by single-chain variable fragment (scFv) antibody that is linked to a superfolder-GFP (sfGFP) followed by TET1cd. sgRNA (FWAg4) was designed to target the ZF108 binding sequence in the *FWA* promoter. Similar to the ZF108-TET1cd experiment, successful demethylation of the *FWA* promoter was achieved resulting in the late-flowering phenotype in transgenic plants. Heritability of the DNA methylation patterns in plants makes this approach very attractive for creation of novel epialleles with agriculturally important traits, while targeted reactivation of transposons can bring additional variation at the genomic level for crop improvement [108, 109].

7 Future Perspectives

The use of SSNs for targeted disruption, insertion, or replacement of selected locus/loci has opened the possibility for precise, fast, and efficient genome editing in crops. These improvements are in turn accelerating both functional and applied genomic studies [110]. The polyploid nature of many crop genomes complicates use of this technology for fast generation of null mutants, but at the same time provides ample raw material for gene expression modulation. Therefore, efforts need to be directed at increasing the efficiency of the SSNs either by improving transformation processes or by better designed GE constructs with improved cloning efficiency that do not compromise target specificity. While the end goal of applying GE technology to crops is improved yield in changing environmental conditions, the widespread cultivation of GE plants can be hampered by regulatory restrictions. Currently, GE plants fall under the Plant Protection Act enforced by USDA [111]. GE cultivars can freely enter into the commercial market if they do not possess any ecological or food safety risks when compared to the same crop developed through conventional breeding techniques. In Canada, both the Canadian Food Inspection Agency (CFIA) and Health Canada regulate only the end product, regardless of the methods used to generate the germplasm [112]. Regulations covering plants with novel trait (PNT) are triggered only when a crop expresses a specific trait with a difference of 20–30% as compared to conventional varieties. In both the USA and Canada, science-based approaches to safely and efficiently regulate novel agricultural products has allowed for continuous introduction of innovative techniques into crops development. However, Europe's highest court ruled on July 25, 2018 that GE crops fall under stringent GM regulations within the European Union (https://www.nature.com/articles/d41586-018-05814-6). Although random mutagenesis techniques developed before 2001 are exempt from this regulation, new varieties created using modern mutagenesis techniques developed after this year, including GE fall under the GM directive.

CRISPR/Cas technology is currently expected to advance more rapidly than ZFNs and TALENs [113]. Simplicity in sgRNA design, relatively easy multiplex targeting, tolerance to DNA methylation and an open access policy within the CRISPR research community are some of the benefits of the CRISPR/Cas technology over the first generation of SSNs. CRISPR plasmids are freely available from the nonprofit repository (e.g., Addgene), and a range of web tools have been developed for selecting gRNA sequences and predicting their specificity (CRISPR-P, CRISPR-PLANT, Cas-OFFinder, etc.) [41]. While the use of this tool in plants opens immense possibilities for plant breeders, the application of CRISPR/Cas9 technology for GE in eukaryotes is now covered by intellectual property awarded in a recent hearing (September 10th, 2018) of the US Court of Appeals for the Federal Circuit to the Broad Institute of MIT and Harvard in Cambridge, Massachusetts (https://www.nature.com/articles/d41586-018-06656-y). It is therefore safe to assume that any new variety developed by CRISPR/Cas9 technology will be subjected to a licencing fee from the Broad Institute. Alternatives to Cas9 (e.g., Cas12a), which are not currently covered by IP protection, are being investigated for their applicability for GE in plants and may grow in popularity over time.

References

1. Bibikova M, Beumer K, Trautman JK et al (2003) Enhancing gene targeting with designed zinc finger nucleases. Science 300:764
2. Puchta H (2002) Gene replacement by homologous recombination in plants. Plant Mol Biol 48:173–182
3. Sakuma T, Woltjen K (2014) Nuclease-mediated genome editing: at the front-line of functional genomics technology. Develop Growth Differ 56:2–13
4. Zhang Y, Zhang F, Li X, Baller JA et al (2013) Transcription activator-like effector nucleases enable efficient plant genome engineering. Plant Physiol 161:20–27
5. Terada R, Urawa H, Inagaki Y et al (2002) Efficient gene targeting by homologous recombination in rice. Nat Biotechnol 20:1030–1034
6. Hanin M, Volrath S, Bogucki A et al (2001) Gene targeting in Arabidopsis. Plant J 28:671–677
7. Voytas DF (2013) Plant genome engineering with sequence-specific nucleases. Annu Rev Plant Biol 64:327–350
8. Choulika A, Perrin A, Dujon B et al (1995) Induction of homologous recombination in mammalian chromosomes by using the I-SceI system of Saccharomyces cerevisiae. Mol Cell Biol 15:1968–1973
9. Cohen-Tannoudji M, Robine S, Choulika A et al (1998) I-SceI-induced gene replacement at a natural locus in embryonic stem cells. Mol Cell Biol 18:1444–1448
10. Puchta H, Dujon B, Hohn B (1993) Homologous recombination in plant cells is enhanced by in vivo induction of double strand breaks into DNA by a site-specific endonuclease. Nucleic Acids Res 21:5034–5040
11. Puchta H, Dujon B, Hohn B (1996) Two different but related mechanisms are used in plants for the repair of genomic double-strand breaks by homologous recombination. Proc Natl Acad Sci U S A 93:5055–5060
12. Gorbunova V, Levy AA (1997) Non-homologous DNA end joining in plant cells is associated with deletions and filler DNA insertions. Nucleic Acids Res 25:4650–4657
13. Mahfouz MM, Piatek A, Stewart CN Jr (2014) Genome engineering via TALENs and CRISPR/Cas9 systems: challenges and perspectives. Plant Biotechnol J 12:1006–1014
14. Osakabe Y, Osakabe K (2015) Genome editing with engineered nucleases in plants. Plant Cell Physiol 56:389–400
15. Heidmann I, de Lange B, Lambalk J et al (2011) Efficient sweet pepper transformation mediated by the BABY BOOM transcription factor. Plant Cell Rep 30:1107–1115
16. Lowe K, Wu E, Wang N et al (2016) Morphogenic regulators Baby boom and Wuschel improve monocot transformation. Plant Cell 28:1998–2015
17. Gaj T, Gersbach CA, Barbas CF 3rd (2013) ZFN, TALEN, and CRISPR/Cas-based methods for genome engineering. Trends Biotechnol 31:397–405
18. Enuameh MS, Asriyan Y, Richards A et al (2013) Global analysis of Drosophila Cys(2)-His(2) zinc finger proteins reveals a multitude of novel recognition motifs and binding determinants. Genome Res 23:928–940
19. Persikov AV, Wetzel JL, Rowland EF et al (2015) A systematic survey of the Cys2His2 zinc finger DNA-binding landscape. Nucleic Acids Res 43:1965–1984
20. Orlando SJ, Santiago Y, DeKelver RC et al (2010) Zinc-finger nuclease-driven targeted integration into mammalian genomes using donors with limited chromosomal homology. Nucleic Acids Res 38:e152
21. Shukla VK, Doyon Y, Miller JC et al (2009) Precise genome modification in the crop species Zea mays using zinc-finger nucleases. Nature 459:437–441
22. Curtin SJ, Zhang F, Sander JD et al (2011) Targeted mutagenesis of duplicated genes in soybean with zinc-finger nucleases. Plant Physiol 156:466–473
23. Ainley WM, Sastry-Dent L, Welter ME et al (2013) Trait stacking via targeted genome editing. Plant Biotechnol J 11:1126–1134
24. Eudes A, Liang Y, Mitra P et al (2014) Lignin bioengineering. Curr Opin Biotechnol 26:189–198

25. Weeks DP (2017) Gene editing in polyploid crops: wheat, camelina, canola, potato, cotton, peanut, sugar cane, and citrus. Prog Mol Biol Transl Sci 149:65–80
26. Christian M, Cermak T, Doyle EL et al (2010) Targeting DNA double-strand breaks with TAL effector nucleases. Genetics 186:757–761
27. Bogdanove AJ, Schornack S, Lahaye T (2010) TAL effectors: finding plant genes for disease and defense. Curr Opin Plant Biol 13:394–401
28. Li T, Huang S, Jiang WZ et al (2011) TAL nucleases (TALNs): hybrid proteins composed of TAL effectors and FokI DNA-cleavage domain. Nucleic Acids Res 39:359–372
29. Boch J, Scholze H, Schornack S et al (2009) Breaking the code of DNA binding specificity of TAL-type III effectors. Science 326:1509–1512
30. Moscou MJ, Bogdanove AJ (2009) A simple cipher governs DNA recognition by TAL effectors. Science 326:1501
31. Haun W, Coffman A, Clasen BM et al (2014) Improved soybean oil quality by targeted mutagenesis of the fatty acid desaturase 2 gene family. Plant Biotechnol J 12:934–940
32. Wang Y, Cheng X, Shan Q et al (2014) Simultaneous editing of three homoeoalleles in hexaploid bread wheat confers heritable resistance to powdery mildew. Nat Biotechnol 32:947–951
33. Wendt T, Holm P, Starker C et al (2013) TAL effector nucleases induce mutations at a pre-selected location in the genome of primary barley transformants. Plant Mol Biol 83:279–285
34. Lor VS, Starker CG, Voytas DF et al (2014) Targeted mutagenesis of the tomato PROCERA gene using transcription activator-like effector nucleases. Plant Physiol 166:1288–1291
35. Clasen BM, Stoddard TJ, Luo S et al (2015) Improving cold storage and processing traits in potato through targeted gene knockout. Plant Biotechnol J 2016(14):169–176
36. Ricroch A, Clairand P, Harwood W (2017) Use of CRISPR systems in plant genome editing: toward new opportunities in agriculture. Emerg Top Life Sci 1:169–182
37. Jinek M, Chylinski K, Fonfara I et al (2012) A programmable dual-RNA-guided DNA endonuclease in adaptive bacterial immunity. Science 337:816–821
38. Langner T, Kamoun S, Belhaj K (2018) CRISPR crops: plant genome editing toward disease resistance. Annu Rev Phytopathol 56:479–512
39. Hsu PD, Scott DA, Weinstein JA et al (2013) DNA targeting specificity of RNA-guided Cas9 nucleases. Nat Biotechnol 31:827–832
40. Qi Lei S, Larson Matthew H, Gilbert Luke A et al (2013) Repurposing CRISPR as an RNA-guided platform for sequence-specific control of gene expression. Cell 152:1173–1183
41. Bortesi L, Fischer R (2015) The CRISPR/Cas9 system for plant genome editing and beyond. Biotechnol Adv 33:41–52
42. Shmakov S, Smargon A, Scott D et al (2017) Diversity and evolution of class 2 CRISPR–Cas systems. Nat Rev Microbiol 15:169–182
43. Zetsche B, Gootenberg Jonathan S et al (2015) Cpf1 is a single RNA-guided endonuclease of a class 2 CRISPR-Cas system. Cell 163:759–771
44. Tang X, Lowder LG, Zhang T et al (2017) A CRISPR–Cpf1 system for efficient genome editing and transcriptional repression in plants. Nat Plants 3:17018
45. Wang M, Mao Y, Lu Y et al (2017) Multiplex gene editing in rice using the CRISPR-Cpf1 system. Mol Plant 10:1011–1013
46. Jaganathan D, Ramasamy K, Sellamuthu G et al (2018) CRISPR for crop improvement: an update review. Front Plant Sci 9:985
47. Li T, Liu B, Spalding MH et al (2012) High-efficiency TALEN-based gene editing produces disease-resistant rice. Nat Biotechnol 30:390–392
48. Wang F, Wang C, Liu P et al (2016) Enhanced rice blast resistance by CRISPR/Cas9-targeted mutagenesis of the ERF transcription factor gene OsERF922. PLoS One 11:e0154027
49. Blanvillain-Baufume S, Reschke M, Sole M et al (2017) Targeted promoter editing for rice resistance to Xanthomonas oryzae pv. oryzae reveals differential activities for SWEET14-inducing TAL effectors. Plant Biotechnol J 15:306–317
50. Zhou H, Liu B, Weeks DP et al (2014) Large chromosomal deletions and heritable small genetic changes induced by CRISPR/Cas9 in rice. Nucleic Acids Res 42:10903–10914
51. Li J, Meng XB, Zong Y et al (2016) Gene replacements and insertions in rice by intron targeting using CRISPR-Cas9. Nat Plants 2:16139
52. Wang MG, Lu YM, Botella JR et al (2017) Gene targeting by homology-directed repair in rice using a geminivirus-based CRISPR/Cas9 system. Mol Plant 10:1007–1010

53. Nekrasov V, Wang C, Win J et al (2017) Rapid generation of a transgene-free powdery mildew resistant tomato by genome deletion. Sci Rep 7:482
54. de Toledo Thomazella DP, Brail Q, Dahlbeck D, et al (2016) CRISPR-Cas9 mediated mutagenesis of a DMR6 ortholog in tomato confers broad-spectrum disease resistance. bioRxiv 064824
55. Chandrasekaran J, Brumin M, Wolf D et al (2016) Development of broad virus resistance in non-transgenic cucumber using CRISPR/Cas9 technology. Mol Plant Pathol 17:1140–1153
56. Lemmon ZH, Reem NT, Dalrymple J et al (2018) Rapid improvement of domestication traits in an orphan crop by genome editing. Nat Plants 4:766–770
57. Zsogon A, Cermak T, Naves ER et al (2018) De novo domestication of wild tomato using genome editing. Nat Biotechnol 36:1211–1216
58. Li TD, Yang XP, Yu Y et al (2018) Domestication of wild tomato is accelerated by genome editing. Nat Biotechnol 36:1160–1163
59. Cermak T, Baltes NJ, Cegan R et al (2015) High-frequency, precise modification of the tomato genome. Genome Biol 16:232
60. Sun YW, Zhang X, Wu CY et al (2016) Engineering herbicide-resistant rice plants through CRISPR/Cas9-mediated homologous recombination of acetolactate synthase. Mol Plant 9:628–631
61. Svitashev S, Young JK, Schwartz C et al (2015) Targeted mutagenesis, precise gene editing, and site-specific gene insertion in maize using Cas9 and guide RNA. Plant Physiol 169:931–945
62. Shimatani Z, Nishizawa-Yokoi A et al (2014) Positive-negative-selection-mediated gene targeting in rice. Front Plant Sci 5:748
63. Fu Y, Foden JA, Khayter C et al (2013) High-frequency off-target mutagenesis induced by CRISPR-Cas nucleases in human cells. Nat Biotechnol 31:822–826
64. Zhang Q, Xing HL, Wang ZP et al (2018) Potential high-frequency off-target mutagenesis induced by CRISPR/Cas9 in Arabidopsis and its prevention. Plant Mol Biol 96:445–456
65. Fujii W, Kawasaki K, Sugiura K et al (2013) Efficient generation of large-scale genome-modified mice using gRNA and CAS9 endonuclease. Nucleic Acids Res 41:e187
66. Pattanayak V, Lin S, Guilinger JP et al (2013) High-throughput profiling of off-target DNA cleavage reveals RNA-programmed Cas9 nuclease specificity. Nat Biotechnol 31:839–843
67. Hahn F, Nekrasov V (2018) CRISPR/Cas precision: do we need to worry about off-targeting in plants? Plant Cell Rep 38:437–441
68. Mao Y, Zhang H, Xu N et al (2013) Application of the CRISPR-Cas system for efficient genome engineering in plants. Mol Plant 6:2008–2011
69. Xie K, Zhang J, Yang Y (2014) Genome-wide prediction of highly specific guide RNA spacers for CRISPR–Cas9-mediated genome editing in model plants and major crops. Mol Plant 7:923–926
70. Kang BC, Yun JY, Kim ST et al (2018) Precision genome engineering through adenine base editing in plants. Nat Plants 4:427–431
71. Li C, Zong Y, Wang YP et al (2018) Expanded base editing in rice and wheat using a Cas9-adenosine deaminase fusion. Genome Biol 19:59
72. Zong Y, Wang YP, Li C et al (2017) Precise base editing in rice, wheat and maize with a Cas9-cytidine deaminase fusion. Nat Biotechnol 35:438–440
73. Shimatani Z, Kashojiya S, Takayama M et al (2017) Targeted base editing in rice and tomato using a CRISPR-Cas9 cytidine deaminase fusion. Nat Biotechnol 35:441–443
74. Lowder LG, Zhang DW, Baltes NJ et al (2015) A CRISPR/Cas9 toolbox for multiplexed plant genome editing and transcriptional regulation. Plant Physiol 169:971–985
75. Piatek A, Ali Z, Baazim H et al (2015) RNA-guided transcriptional regulation in planta via synthetic dCas9-based transcription factors. Plant Biotechnol J 13:578–589
76. Lowder LG, Zhou JP, Zhang YX et al (2018) Robust transcriptional activation in plants using multiplexed CRISPR-Act2.0 and mTALE-act systems. Mol Plant 11:245–256
77. Woo JW, Kim J, Kwon SI et al (2015) DNA-free genome editing in plants with preassembled CRISPR-Cas9 ribonucleoproteins. Nat Biotechnol 33:1162–1164
78. Zhang Y, Liang Z, Zong Y et al (2016) Efficient and transgene-free genome editing in wheat through transient expression of CRISPR/Cas9 DNA or RNA. Nat Commun 7:12617
79. Liang Z, Chen K, Li T et al (2017) Efficient DNA-free genome editing of bread wheat using CRISPR/Cas9 ribonucleoprotein complexes. Nat Commun 8:14261

80. Svitashev S, Schwartz C, Lenderts B et al (2016) Genome editing in maize directed by CRISPR-Cas9 ribonucleoprotein complexes. Nat Commun 7:13274
81. Andersson M, Turesson H, Olsson N et al (2018) Genome editing in potato via CRISPR-Cas9 ribonucleoprotein delivery. Physiol Plant 164:378–384
82. Malnoy M, Viola R, Jung MH et al (2016) DNA-free genetically edited grapevine and apple protoplast using CRISPR/Cas9 ribonucleoproteins. Front Plant Sci 7:1904
83. Bilichak A, Luu J, Eudes F (2015) Intracellular delivery of fluorescent protein into viable wheat microspores using cationic peptides. Front Plant Sci 6:666
84. Meinke DW, Franzmann LH, Nickle TC et al (1994) Leafy cotyledon mutants of Arabidopsis. Plant Cell 6:1049–1064
85. West M, Yee KM, Danao J et al (1994) LEAFY COTYLEDON1 is an essential regulator of late embryogenesis and cotyledon identity in Arabidopsis. Plant Cell 6:1731–1745
86. Ogas J, Kaufmann S, Henderson J et al (1999) PICKLE is a CHD3 chromatin-remodeling factor that regulates the transition from embryonic to vegetative development in Arabidopsis. Proc Natl Acad Sci USA 96:13839–13844
87. Hecht V, Vielle-Calzada JP, Hartog MV et al (2001) The Arabidopsis SOMATIC EMBRYOGENESIS RECEPTOR KINASE 1 gene is expressed in developing ovules and embryos and enhances embryogenic competence in culture. Plant Physiol 127:803–816
88. Heck GR, Perry SE, Nichols KW et al (1995) Agl15, a Mads domain protein expressed in developing embryos. Plant Cell 7:1271–1282
89. Gazzarrini S, Tsuchiya Y, Lumba S et al (2004) The transcription factor FUSCA3 controls developmental timing in Arabidopsis through the hormones gibberellin and abscisic acid. Dev Cell 7:373–385
90. Elhiti M, Tahir M, Gulden RH et al (2010) Modulation of embryo-forming capacity in culture through the expression of Brassica genes involved in the regulation of the shoot apical meristem. J Exp Bot 61:4069–4085
91. Zuo JR, Niu QW, Frugis G et al (2002) The WUSCHEL gene promotes vegetative-to-embryonic transition in Arabidopsis. Plant J 30:349–359
92. Boutilier K, Offringa R, Sharma VK et al (2002) Ectopic expression of BABY BOOM triggers a conversion from vegetative to embryonic growth. Plant Cell 14:1737–1749
93. Horstman A, Li MF, Heidmann I et al (2017) The BABY BOOM transcription factor activates the LEC1-ABI3-FUS3-LEC2 network to induce somatic embryogenesis. Plant Physiol 175:848–857
94. Lowe K, La Rota M, Hoerster G et al (2018) Rapid genotype "independent" Zea mays L. (maize) transformation via direct somatic embryogenesis. In Vitro Cell Dev Biol Plant 54:240–252
95. Mookkan M, Nelson-Vasilchik K, Hague J et al (2017) Selectable marker independent transformation of recalcitrant maize inbred B73 and sorghum P898012 mediated by morphogenic regulators BABY BOOM and WUSCHEL2. Plant Cell Rep 36:1477–1491
96. Khanday I, Skinner D, Yang B et al (2019) A male-expressed rice embryogenic trigger redirected for asexual propagation through seeds. Nature 565:91–95
97. Bilichak A, Luu J, Jiang F et al (2018) Identification of BABY BOOM homolog in bread wheat. Agri Gene 7:43–51
98. Ou X, Zhang Y, Xu C et al (2012) Transgenerational inheritance of modified DNA methylation patterns and enhanced tolerance induced by heavy metal stress in rice (Oryza sativa L.). PLoS One 7:e41143
99. Abdelnoor RV, Yule R, Elo A et al (2003) Substoichiometric shifting in the plant mitochondrial genome is influenced by a gene homologous to MutS. Proc Natl Acad Sci USA 100:5968–5973
100. Xu YZ, Arrieta-Montiel MP, Virdi KS et al (2011) MutS HOMOLOG1 is a nucleoid protein that alters mitochondrial and plastid properties and plant response to high light. Plant Cell 23:3428–3441
101. Santamaria RD, Shao MR, Wang GM et al (2014) MSH1-induced non-genetic variation provides a source of phenotypic diversity in Sorghum bicolor. PLoS One 9:e108407
102. Virdi KS, Laurie JD, Xu YZ et al (2015) Arabidopsis MSH1 mutation alters the epigenome and produces heritable changes in plant growth. Nat Commun 6:6386
103. Yang XD, Kundariya H, Xu YZ et al (2015) MutS HOMOLOG1-derived epigenetic breeding potential in tomato. Plant Physiol 168:222–U390
104. Raju SKK, Shao MR, Sanchez R et al (2018) An epigenetic breeding system in soybean for increased yield and stability. Plant Biotechnol J 16:1836–1847

105. Gallego-Bartolome J, Gardiner J, Liu WL et al (2018) Targeted DNA demethylation of the Arabidopsis genome using the human TET1 catalytic domain. Proc Natl Acad Sci USA 115:E2125–E2134
106. Papikian A, Liu WL, Gallego-Bartolome J et al (2019) Site-specific manipulation of Arabidopsis loci using CRISPR-Cas9 SunTag systems. Nat Commun 10:729
107. Morita S, Noguchi H, Horii T et al (2016) Targeted DNA demethylation in vivo using dCas9-peptide repeat and scFv-TET1 catalytic domain fusions. Nat Biotechnol 34:1060–1065
108. Hofmeister BT, Lee K, Rohr NA et al (2017) Stable inheritance of DNA methylation allows creation of epigenotype maps and the study of epiallele inheritance patterns in the absence of genetic variation. Genome Biol 18:155
109. Niederhuth CE, Schmitz RJ (2014) Covering your bases: inheritance of DNA methylation in plant genomes. Mol Plant 7:472–480
110. Petolino JF (2015) Genome editing in plants via designed zinc finger nucleases. In Vitro Cell Dev Biol Plant 51:1–8
111. Duensing N, Sprink T, Parrott WA et al (2018) Novel features and considerations for ERA and regulation of crops produced by genome editing. Front Bioeng Biotechnol 6:79
112. Smyth SJ (2017) Canadian regulatory perspectives on genome engineered crops. GM Crops Food 8:35–43
113. Belhaj K, Chaparro-Garcia A, Kamoun S et al (2015) Editing plant genomes with CRISPR/Cas9. Curr Opin Biotechnol 32:76–84

Chapter 15

CRISPR/Cas9-Mediated Targeted Mutagenesis in Wheat Doubled Haploids

Alison M. R. Ferrie, Pankaj Bhowmik, Nandhakishore Rajagopalan, and Sateesh Kagale

Abstract

CRISPR/Cas9-based genome editing technology has the potential to revolutionize agriculture, but many plant species and/or genotypes are recalcitrant to conventional transformation methods. Additionally, the long generation time of crop plants poses a significant obstacle to effective application of gene editing technology, as it takes a long time to produce modified homozygous genotypes. The haploid single-celled microspores are an attractive target for gene editing experiments, as they enable generation of homozygous doubled haploid mutants in one generation. Here, we describe optimized methods for genome editing of haploid wheat microspores and production of doubled haploid plants by microspore culture.

Key words Wheat, Microspores, Genome editing, CRISPR, gRNA, Cas9, In vitro nuclease assay, Neon transfection system, Doubled haploids, Microspore culture

1 Introduction

Wheat (*Triticum aestivum* L.) is an important staple food source globally, providing 20–25% of the daily calorific and protein intake by world population (www.fao.org/faostat). Wheat has a large allohexaploid genome of 16 GB with approximately 85% repetitive DNA elements [1]. Due to its complex genome and difficulties in genetic transformation, wheat has lagged behind other cereals in the application of innovative molecular techniques of genetic engineering. The presence of multiple copies of each gene with high levels of sequence identity in wheat makes its genetic manipulation very tricky.

CRISPR/Cas9 (Clustered, Regularly Interspersed, Palindromic Repeats/CRISPR-associated endonuclease 9)-based genome editing has emerged as a powerful technique that can unlock the genetic potential of the complex wheat genome. With the availability of advanced multiplex genome editing tools [2, 3], simultaneous modification of multiple copies of a gene or gene family has

Luis M. Vaschetto (ed.), *Cereal Genomics: Methods and Protocols*, Methods in Molecular Biology, vol. 2072, https://doi.org/10.1007/978-1-4939-9865-4_15,

become possible. The recent success in application of CRISPR/Cas9-based genome editing to develop wheat lines with hypoimmunogenic gluten [4], increased grain size and weight [5, 6] and disease resistance [7] has demonstrated the utility of CRISPR/Cas9 for breeding improved wheat. However, most of the current gene editing approaches in wheat use the conventional immature embryo-based genetic transformation methods, such as *Agrobacterium*-mediated transformation [6, 8] or biolistic bombardment [9, 10], which are less efficient, labor-intensive, and time-consuming. Apart from a few genotypes (Fielder and Bobwhite), most of the wheat cultivars are recalcitrant to conventional transformation procedures [11]. Also, it can take several generations to obtain homozygous lines of conventionally transformed plants.

Doubled haploidy, a widely used technique in crop breeding programs [12, 13], can help to accelerate the process of genome editing in wheat and other crops. Microspore embryogenesis is a simple and affordable technique for the production of doubled haploid plants. Microspores are also a desirable explant for genome editing because of their relative ease and speed of isolation, genetic and physiological uniformity, reduced complexity of the genome (the number of alleles to be edited is reduced by half due to the haploid genome), the potential for generation of homozygous doubled haploid plants in one generation, and the absence of chimerism. We recently demonstrated the feasibility of combining microspore-based doubled haploid technology and CRISPR/Cas9-based gene editing for targeted mutagenesis in wheat [14]. We optimized a number of factors that may affect the delivery of CRISPR/Cas9 reagents into microspores, including pulsing voltage for electroporation, the composition of the electroporation buffer, number of microspores and the amount of plasmid DNA. In this book chapter, we describe in detail a step-by-step protocol for CRISPR/Cas9-mediated targeted mutagenesis in wheat microspores followed by the production of doubled haploid plants.

The haploid mutagenesis approach described in this chapter represents an effective and efficient strategy to accelerate the pace of gene discovery and crop breeding. The novelty of this system lies in the combination of the CRISPR/Cas9-based gene editing and the microspore-derived doubled haploid technologies. This method is applicable to other crops provided reliable protocols for the production of doubled haploid plants through microspore culture are established. Future research aimed at optimizing an efficient transfection system for directly delivering ribonucleoprotein complexes into microspores would establish the desired DNA-free haploid genome editing system.

2 Materials

2.1 Equipment

1. Neon transfection system.
2. PCR thermocycler.
3. Agilent 2100 Bioanalyzer.
4. Growth cabinets.
5. Laminar air flow hood.
6. Refrigerated centrifuge, benchtop.
7. Incubators, tissue culture room.
8. Microscope (e.g., PhotoZoom inverted microscope, fluorescence microscope).
9. General tissue culture equipment.

2.2 Supplies and Reagents

1. 1.5 mL microcentrifuge tubes.
2. Glass bottom microwell dishes (Mat Tek Corporation).
3. Nuclease-free water.
4. *Streptococcus pyogenes* Cas9 nuclease (Prepared in-house or New England BioLabs).
5. 10× reaction buffer (10× NEBuffer™ 3.1, New England BioLabs).
6. gRNA.
7. Proteinase K.
8. Neon™ transfection system 100 μL Kit (ThermoFisher Scientific).
9. Microspore culture media—NPB-99 medium with glutathione.
10. Taq DNA polymerase (QIAGEN).
11. QIAquick PCR Purification Kit (QIAGEN).
12. QIAquick Gel Extraction Kit (QIAGEN).
13. High Fidelity *Sac I* restriction enzyme (NEB) supplied with CutSmart® Buffer.
14. TOPO™ TA Cloning™ Kit for Sequencing (Thermo Fisher Scientific).
15. DNeasy Plant Mini Kit (QIAGEN).
16. RNaseZap™ (ThermoFisher Scientific).
17. 8″ AZ pots using Sunshine Mix #4.

2.3 Buffers

1. Cas9 buffer (20 mM HEPES pH 7.5, 200 mM KCl, 10 mM $MgCl_2$, 0.5 mM TCEP). Mix together 8 mL of 1 M HEPES pH 7.5, 40 mL of 2 M KCl, 4 mL of 1 M $MgCl_2$, 0.4 mL of

0.5 M TCEP, adjust final volume to 400 mL with water, sterilize by passing through a 0.22 μM filter. Store at 4 °C for up to a week.

2. Composition of media used in wheat microspore culture and plantlet regeneration (*see* Table 1).

3 Methods

The standardized protocol for targeted mutagenesis in wheat doubled haploids can be divided into five sequential steps: (1) design of effective guide RNA (gRNA), (2) validation of gRNA efficacy, (3) microspore isolation, (4) delivery of CRISPR/Cas9 components into microspores using the Neon electroporation system, and (5) microspore culture and chromosome doubling.

3.1 Design of Effective gRNA

Models for predicting on-target activity and off-target potential of CRISPR/Cas9 using empirical evidence have been developed [15, 16]. We applied these to wheat and designed a bioinformatics portal (WheatCRISPR; Cram et al., submitted; https://crispr.bioinfo.nrc.ca/WheatCrispr/) to design gRNAs for CRISPR editing. WheatCRISPR allows researchers to browse all possible gRNAs targeting a gene of interest and select effective gRNAs based on the predicted high on-target and low off-target activity scores, as well as other characteristics such as position within the targeted gene.

1. The input for WheatCRISPR is either the target gene identifier (IWGSC annotation v1.1) or the DNA sequence of the target gene.
2. WheatCRISPR allows users to design gRNA for the coding or the promoter (2 kbp immediately upstream of the gene) regions of the target gene. The user can also choose to target a specific gene or all homoeologous copies of the gene.
3. The outputs include a gRNA table, gRNA plot, gene plot and off-targets table for a given gene. The gRNA table and plot summarize the on-target (rs2, rule-set 2) and the off-target (CFD, cutting frequency determination) activity scores for each of the four genomic regions: coding, promoter, other genic, and intergenic. The gene plot displays the physical location of the gRNAs against the gene models. The off-target table displays all off-target hits for the selected gRNA. The gRNAs are ranked based on an overall score, which is a weighted average of the rs2 and maximum CFD scores.
4. Effective gRNAs are selected based on high overall score, high on-target activity (rs2 score) and low predicted off-target potential (low CFD scores).

Table 1
Composition of media used in wheat microspore culture and plantlet regeneration

	Media			
Components (mg/L)	**FHG-2 solution**	**NPB-99⁺ medium**	**0.3 M Mannitol solution**	**B5-5 regeneration**
KNO_3	1900.00	1415.0		2500.0
$(NH_4)NO_3$	165.00			
$(NH_4)_2SO_4$		232.0		134.0
KH_2PO_4	170.00	200.0		
$NaH_2PO_4{\cdot}H_2O$				150.0
$CaCl_2{\cdot}2H_2O$	440.00	83.0		150.0
$MgSO_4{\cdot}7H_2O$	370.00	93.0		250.0
$FeSO_4{\cdot}7H_2O$	27.85	27.8		27.85
Na_2EDTA	37.25	37.3		37.25
$MES{\cdot}H_2O$	1950.00		500.0	1000.0
KI		0.4		0.75
$MnSO_4{\cdot}H_2O$		5.0		10.0
H_3BO_3		5.0		3.0
$ZnSO_4{\cdot}7H_2O$		5.0		2.0
$CoCl_2{\cdot}6H_2O$		0.0125		0.025
$CuSO_4{\cdot}5H_2O$		0.0125		0.025
$Na_2MoO_4{\cdot}2H_2O$		0.0125		0.25
Thiamine HCl		5.0		10.0
Pyridoxine HCl		0.5		1.0
Nicotinic acid		0.5		1.0
myo-Inositol		50.0		100.0
Glutamine		500.0		
Glutathione		0.61		
Larcoll		40.0		
Sucrose				20,000.0
Maltose		90,000.0		
Mannitol	72,870.00		54,650.0	
2,4-D		0.2		
Kinetin		0.2		1.0
PAA		1.0		
IAA				1.0
Cefotaxime		100.0		
pH (final)	6.5	6.5	6.5	5.8
Difco agar				7000.0

3.2 Validation of gRNA Efficacy

We use an in vitro nuclease assay (described below) to validate the efficacy of selected gRNAs (exemplified in ref. 17).

3.2.1 Sample Dilutions

1. Dilute gRNA and Cas9 protein to 300 nM using nuclease-free water and Cas9 buffer, respectively.
2. Polymerase chain reaction (PCR)-amplify DNA of the target gene region encompassing the gRNAs and dilute it to 30 nM using nuclease-free water.

3.2.2 Nuclease Assay Setup

1. The three main components gRNA, Cas9 nuclease and target DNA must be present at a final molar ratio of 10:10:1 in a total assay volume of 30 μL.
2. Mix the assay reagents in the following order:
3. Mix and incubate at room temperature (21 °C) for 10 min.

Reagent	Assay reaction (μL)	Control 1 (no Cas9) (μL)	Control 2 (no gRNA) (μL)
Nuclease-free water	18	21	21
10× reaction buffer	3	3	3
300 nM gRNA	3	3	0
300 nM Cas9 nuclease	3	0	3

4. Add 3 μL of 30 nM target DNA to assay and control reactions.
5. Mix and incubate at 37 °C for 15 min. Longer incubation times for the reaction at 37 °C can lead to complete digestion of the target DNA and formation of products.
6. Stop reaction by adding 1 μL of Proteinase K to each tube.
7. Mix and incubate at room temperature (21 °C) for 10 min.
8. Analyze assay reaction products on an agarose gel or using a BioAnalyzer (Agilent).
9. The band sizes of the cleaved product of the target DNA using different gRNAs and Cas9 endonuclease will reveal the effectiveness of different gRNAs, as exemplified in Figure 3 of Rajagopalan et al. [17].
10. Some useful tips for in vitro nuclease assay are provided in **Note 1**.

3.3 Microspore Isolation

3.3.1 Growing Donor Plants: Spring Wheat

1. Four seeds are planted into 8″ AZ pots using Sunshine Mix #4 with ~4.5 g of slow release fertilizer, 14-13-13. Growing conditions are set at 20/18 °C, 18 h photoperiod, and 350–420 μE light intensity.
2. After germination thin the plants to 3 per pot. Use dilute 0.22 g/L 20-20-20 fertilizer when watering.

3.3.2 Growing Donor Plants: Winter Wheat

1. Four seeds are planted into 8″ AZ pots using Sunshine Mix #4 with ~4.5 g of slow release fertilizer, 14-13-13. Growing conditions are initially set at 20/18 °C, 18 h photoperiod, and 350–420 μE light intensity.
2. After germination thin the plants to 3 per pot. Alternatively, seeds can be planted into root trainers containing Sunshine Mix #4.
3. After 2 weeks, transfer the plants in pots or root trainers to 3 °C, 12 h photoperiod, and 300 μE light intensity for a 9-week vernalization treatment.
4. Use dilute 0.22 g/L 20-20-20 fertilizer when watering. At 3 °C the plants require less watering than at 20/18 °C.
5. After vernalization, transfer the plants to 15/12 °C, 18 h photoperiod, and 400–420 μE light intensity. If the plants were started in root trainers transplant them to 8″AZ pots containing Sunshine Mix #4, 3 plants per pot, and add ~4.5 g of slow release fertilizer (14-13-13).

3.3.3 Collecting Spikes for Microspore Culture

1. Before spikes can be collected for experiments, one needs to determine when the majority of the microspores are at the mid–late uninucleate stage (Fig. 1). This can then be correlated to location of the spike in the boot. The location of the spike in the boot may vary from growth cabinet to growth cabinet.
2. Collect spikes at the correct microspore developmental stage as determined previously. Trim off leaves, keeping 2–3 cm of the flag leaf.
3. Place spikes in a flask containing about 200 mL 4 °C water.
4. Store spikes at 4 °C for 21–24 days for spring wheat or 28–30 days for winter wheat before experimentation. Do not let the spikes run out of water.
5. Use 10–12 spikes per experiment. An experiment can be performed with fewer spikes, the results will be fewer experimental plates.

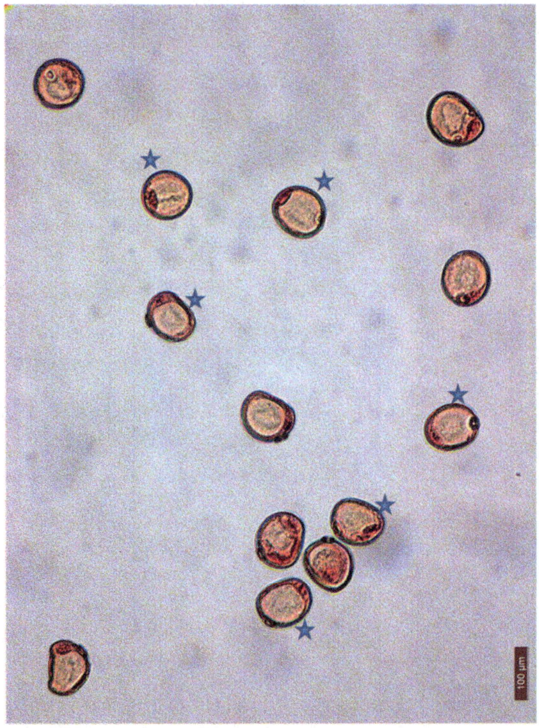

Fig. 1 Microspores at the mid to late uninucleate stage (acetocarmine stain). Stars indicate microspores at optimal stage

3.3.4 Sterilization of Spikes for Microspore Culture

1. Trim off the awns. Place the heads in a sterile 500 mL bottle and sterilize spikes with 500 mL of 10% bleach with Tween 80 (three drops/500 mL) for 3 min.
2. Rinse four times with 500 mL of cold sterile water.

3.3.5 Microspore Isolation

1. Keep the bottle of sterile microspore spikes on ice.
2. Place medium blender cup on ice and fill approximately ½ full with sterile cold FHG-2 extraction buffer (Table 1). Working on a sterile surface, trim off awns and remove the glumes from either side of the spikelet.
3. Remove each spikelet from the rachis and transfer to the blender cup. After two or three spikes are complete, use sterile forceps to submerge the spikelets in the extraction medium. Repeat until all the spikes are used.
4. Blend the spikelets on low speed twice for 10 s each.
5. Pour the blended suspension through a 90 μm filter funnel into a sterile bottle on ice.
6. Rinse the blender cup with cold FHG-2 medium and pour through the filter. Pour cold FHG-2 medium through the filter to give a final volume of 300 mL in the bottle.

7. Pour the filtrate into 6 × 50 mL centrifuge tubes and centrifuge at 200 *g* 4 °C for 5 min.
8. Decant each tube.
9. Pour 20 mL cold FHG-2 medium into one tube. Rinse each tube with the same 20 mL aliquot pooling the microspore solution in the last tube. Repeat with another 20 mL cold FHG-2 medium.
10. Add cold FHG-2 medium to the combined pellets bringing the volume up to 50 mL.
11. Centrifuge at 200 *g* 4 °C for 5 min.
12. Decant and resuspend pellet in 5 mL of cold NPB-99 medium with glutathione, Larcoll, and cefotaxime (NPB99 + GLC; Table 1). Pour microspore suspension into a 15 mL tube. Rinse the 50 mL tube with an additional 5 mL cold NPB99 + GLC and pour into the 15 mL tube. Repeat to a final volume of 15 mL in the 15 mL tube.
13. Centrifuge at 200 *g* 4 °C for 5 min. Decant.
14. Add 10 mL of cold 23% maltose in H_2O. Gently resuspend the microspores. Carefully layer 1 mL of cold NPB99 + GLC over the maltose solution. Centrifuge at 200 *g* 4 °C for 13 min. A distinct band of microspores will form at the interface. Carefully remove the band of microspores and transfer to a clean sterile 15 mL tube.
15. Bring microspores to 6 mL volume with cold NPB99 + GLC and then to 15 mL volume with cold 0.3 M mannitol (Table 1) using the measurements on the tube. Invert mix.
16. Centrifuge at 200 *g* 4 °C for 5 min. Decant.
17. For ease of counting on the hemocytometer bring microspores to 3 or 6 mL volume with cold NPB99 + GLC depending on the pellet size. If the pellet size is ~0.05 mL add 3 mL cold NPB99 + GLC to the microspores but if the pellet is 0.1 mL add 6 mL cold NPB99 + GLC. Using a transfer pipette resuspend the microspores uniformly throughout.
18. Upon attaining a uniform microspore suspension apply 12 μL to one grid of a hemocytometer. Count the microspores in four 1 mm^2 sections (4 × 4—16 squares) of the hemocytometer. Repeat. Average all 8 counts obtaining an average microspore count per quadrant.
19. If the microspores were resuspended in 3 mL NPB99 + GLC multiply the average number of microspores per quadrant by 0.311 to obtain the number of plates for the experiment. If the microspores were resuspended in 6 mL NPB99 + GLC multiply the average number of microspores per quadrant by 2 and then by 0.311 to obtain the number of plates for the experi-

ment. Multiply this number by 1.5 mL per plate to obtain the total volume for the experiment.

20. Trichostatin A (TSA) is a histone deacetylase inhibitor that has been shown to be effective in numerous wheat genotypes. The concentration 0.008 μM TSA is used for cv. AC Nanda [18]. Stock concentrations in DMSO at 50 and 5 μM TSA can be prepared. Depending on the microspore volume working with, TSA is added to the culture medium using the formula C1V1 = C2V2.

3.4 Delivery of CRISPR/Cas9 Components into Haploid Microspores

The optimized protocol for wheat microspore transfection using the Neon system involves electroporation of a minimum of 75,000 cells using 10–20 μg of plasmid DNA and a pulsing voltage of 500 V (three pulses of 20 s each) [14].

3.4.1 Preparation of Microspores for Electroporation

1. Prepare an aliquot of at least 75,000 microspores in 1.5 mL micro centrifuge tube.
2. Remove the NPB99 + GLC media from tube and resuspended the microspores in Neon electroporation buffer (R buffer).
3. Add 20–30 μg of plasmid DNA carrying CRISPR/Cas9 construct to the tube containing microspores and mix by gently flicking the tube and adjust the volume to 100 μL using the R buffer. An ideal CRISPR/Cas9 construct for microspore transfection should consist of constitutive expression cassettes for (1) single guide RNA (sgRNA), (2) Cas9 nuclease (preferably optimised for heterologous expression in wheat), and (3) a fluorescent protein reporter gene (*see* **Note 2**).

3.4.2 Microspore Transfection Using the Neon Electroporation System

Perform electroporation following the protocol described in the Neon transfection manual (https://www.thermofisher.com/order/catalog/product/MPK5000).

1. Set up a Neon® Tube with 3 mL Electrolytic Buffer E2 into the Neon® Pipette Station. Set pulse conditions at 500 Voltage with 20 pulse width and the number of pulses as 3.
2. Collect the microspore–DNA mixture from the microcentrifuge tube using the Neon® Pipette. Avoid air bubbles during pipetting as they cause arcing during electroporation, leading to low or failed transfection.
3. Press Start on the touchscreen.
4. After electroporation slowly remove the Neon® Pipette from the Neon® Pipette Station. Immediately transfer the microspores from the Neon® Tip into the culture plate containing 3 mL of the NPB99 culture medium.
5. Observe transfected microspores under the microscope for viability.

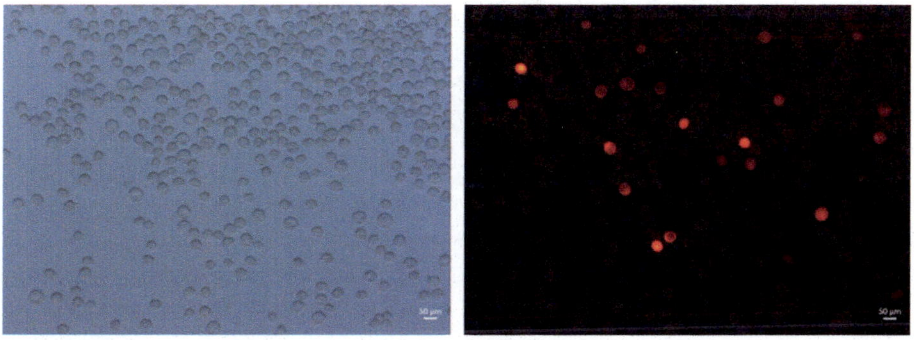

Fig. 2 Successful delivery of a plasmid carrying red fluorescent protein reporter gene cassette into wheat microspores using the Neon electroporation system. The image in the left panel shows cell viability and the right panel shows cells expressing DsRed

6. Confirm CRISPR/Cas9 construct delivery by microscopic imaging of cells for expression of the fluorescent reporter protein. Figure 2 shows an epifluorescent microscopic image of microspores transfected with a red fluorescent protein (DsRed) reporter gene.
7. Verify edits at the expected cut sites through amplicon sequencing.
8. (Optional) Enrich transfected microspores using a fluorescence assisted cell sorter.
9. Dispense 1.5 mL microspore suspension per 35 × 10 mm plate to produce a density of 60,000 microspores per mL.

3.5 Microspore Culture

3.5.1 Ovary Explant Preparation

The developmental stage of the ovary is determined prior to the day of experimentation. Ovaries at the elongated/cleft stage are desired (Fig. 3). The spikes selected for ovaries can be cut fresh the day of experimentation or cut 1–2 weeks prior to use and stored at 4 °C. If storing ovary spikes trim off the leaves keeping 5–7.5 cm of the flag leaf. Loosely wrap ovary spikes in aluminum foil.

3.5.2 Sterilization of Ovary Spikes for Microspore Culture

1. For sterilization, trim awns and place spikes in a sterile 500 mL bottle. Sterilize in 500 mL of 20% bleach with Tween 80 (three drops/500 mL) for 5 min.
2. Rinse four times with 500 mL of cold sterile water.
3. Store the sterilized heads in the bottle at 4 °C until needed.

3.5.3 Addition of Ovaries

1. Aseptically isolate ovaries from the sterile ovary spikes selecting elongated/cleft ovaries (Fig. 3).
2. Place three ovaries into each plate of microspores.
3. Wrap with Parafilm and place plates inside a large petri dish (150 × 25 mm) with an open 60 × 15 mm petri dish filled with water in the center.

Fig. 3 Developmental stage of wheat ovaries desirable for microspore culture experiments

4. Incubate microspores in the dark at 28 °C for 28–35 days. Do not disturb.

3.5.4 Plating Embryoids

1. After 28–35 days incubation at 28 °C count the embryoids using a dissecting microscope.
2. In a laminar flow hood, aseptically remove and discard the ovaries. Pour the embryoids onto solid B5-5 medium (Table 1). If the embryoid count is very high distribute the embryoids over several plates such that there are approximately 100–200 embryoids per plate (100 × 25 mm plates).
3. Draw off the liquid medium with a sterile transfer pipette.
4. Using the transfer pipette or forceps gently distribute the embryoids over the surface of the medium.
5. Seal the plate with Parafilm.
6. Place plates at 22 °C, 16 h photoperiod, and 100 μE light intensity. Shoots should start developing within 1–2 weeks.
7. After 2–3 weeks, developing green shoots can be transferred from the petri plates to Magenta boxes containing B5-5 medium.
8. Once the shoots have a well-established root system, they can be checked for ploidy using a flow cytometer. We have used a Partec Cell Counter Analyzer or the Beckman Coulter CytoFLEX Model A00-1-1102. In either case, young leaf tissue (0.5 cm^2) is chopped and stained in Partec CyStain UV Precise P, passed through a 30 μm filter, and the filtrates are analyzed through multiwell plate or single-tube setups.

9. If the plantlet has spontaneously doubled, then the plantlet can be transferred directly to soil, if no spontaneous doubling has taken place, then a chromosome doubling treatment will be necessary. The frequency of spontaneous doubling varies depending on genotype.
10. Some useful tips for microspore culture are provided in **Note 3**.

3.5.5 Chromosome Doubling

1. The haploid plantlet is removed from the petri plate and the agar is gently removed from the roots.
2. The roots are then submerged in a 0.3% solution of colchicine for 4 h.
3. After 4 h, the roots are rinsed in water and the plantlet is transferred to a soil-less mix (Sunshine Mix #4) and grown in the greenhouse or growth cabinet (at 20/18 °C, 18 h photoperiod). Plantlets are initially covered with plastic cups to maintain a high humidity. This covering is slowly removed (over a period of 7–10 days) as the plants harden.
4. Doubled haploid plants can be grown in the growth cabinet or greenhouse to maturity.

4 Notes

1. ***Useful tips for* in vitro *nuclease assay***
   (a) Prepare a clean working environment by spraying and wiping work surfaces with an RNase decontamination solution RNaseZap™.
   (b) Use certified nuclease-free pipette tips and tubes to set up the reactions.
   (c) Use nuclease-free water for all sample dilutions.
   (d) Either a PCR amplified DNA fragment or a DNA fragment cloned into a plasmid vector could be used as the target DNA for cleavage.
2. ***Enrichment of transfected microspores***

   Microspore transfection efficiency is generally low. The Neon electroporation method provided highest transfection efficiency of only 2.2% [14]. Thus, it would be worthwhile to include an easily detectable fluorescence protein marker on the gene editing construct. The presence of a fluorescent reporter protein will allow sorting of transfected microspores accurately and quantitatively using a fluorescence-activated cell sorter [19, 20].

3. *Useful tips for microspore culture*

(1) Determining the developmental stage of the microspore is critical. This needs to be completed prior to selecting the experimental material. During pretreatment, plant growth continues and microspore culture induction occurs. For example, winter wheat variety AC Radiant [21] microspores have more late uninucleate to early binucleate staged microspores when the spike is 1.5–2.5 cm below the flag leaf. To allow for growth during pretreatment, AC Radiant tillers are cut when the spike in the boot is 2.5–3.5 cm below the flag leaf. For all genotypes, each is initially staged to determine the distance of the head below the flag leaf to obtain mid to late uninucleate microspores and then 1 cm is added to that distance. Therefore: Cutting measurement = staging measurement +1 cm

(2) During pretreatment, spikes may emerge from or split the side of the boot. These spikes are still used.

(3) Spikes for microspores or ovaries can be sterilized at the same time and then stored at 4 °C until needed.

(4) Mix bleach solutions fresh before each experiment.

(5) Items to keep stocked at 4 °C—media (extraction medium, NBP99 + GLC, maltose, mannitol), sterile blender cups, sterile 90 μm funnel filters, sterile 500 mL bottles, and sterile bottles of water (500 mL and/or 1 L).

(6) Work on ice throughout the extraction.

(7) For the microspore (and later ovary) isolation, sterile 100 × 15 mm petri plates or sterile paper towels can be used for a sterile working surface.

(8) Prior to decanting estimate the size of the pellet using the markings on the 15 mL tube. An estimate of pellet size can give you a very rough estimate of the number of plates to expect. For example, a pellet of 0.1 mL will give you roughly 20 ± 5 plates.

(9) Stock concentrations in DMSO at 50 μM and 5 μM TSA can be prepared. Depending on the microspore volume working with, TSA is added to the culture medium using the formula C1V1 = C2V2.

(10) Example calculation for microspores resuspended in 3 mL NPB99 + GLC

(a) $[(41 + 28 + 30 + 37) + (22 + 21 + 26 + 26)] \div 8 = 231 \div 8 = 28.875$ round down. Average microspores per quadrant is 28

(b) $28 \times 0.311 = 8.71$ round down therefore 8 plates for this experiment

(c) 8 plates × 1.5 mL per plate = 12 mL total volume for this experiment

(d) 12 – 3 mL microspore suspension = 9 mL media to add

(e) Calculate the amount of TSA to add based on the concentration of your TSA stock and the total volume for the experiment. C1V1 = C2V2 and add it to your tube. Mix the TSA and NPB99 + GLC.

(f) Add 3 mL microspore suspension to tube. Mix and plate 1.5 mL per 35 × 10 mm plate. This produces a density of 60,000 microspores per mL, 1.5 mL per plate, and approximately 90,000 microspores per plate. Label and number the plates.

(11) Example calculation for microspores resuspended in 6 mL NPB99 + GLC

(a) Calculate the average microspores per quadrant the same as in **step** (**a**) above.

(b) Average microspores per quadrant × 2 × 0.311 = number of plates for the experiment.

(c) Continue with the calculations as above starting at (**c**).

(12) The germinating microspore-derived embryos may develop green or albino shoots. The albino shoots are discarded.

Acknowledgments

The authors acknowledge H.M. Wang, J.L. Enns, K.L. Nelson, J.M. Brost, T.D. Orr, and K. Caswell for their contribution to the doubled haploidy protocol development in wheat and B. Polley and H. Song for their contribution to standardization of the microspore transfection protocol.

References

1. International Wheat Genome Sequencing Consortium (IWGSC); IWGSC RefSeq principal investigators, Appels R, et al (2018) Shifting the limits in wheat research and breeding using a fully annotated reference genome. Science 361:eaar7191
2. Wang W, Pan Q, He F et al (2018) Transgenerational CRISPR-Cas9 activity facilitates multiplex gene editing in allopolyploid wheat. CRISPR J 1:65–74
3. Xie K, Minkenberg B, Yang Y (2015) Boosting CRISPR/Cas9 multiplex editing capability with the endogenous tRNA-processing system. Proc Natl Acad Sci U S A 112:3570–3575
4. Sanchez-Leon S, Gil-Humanes J, Ozuna CV (2017) Low-gluten, non-transgenic wheat engineered with CRISPR/Cas9. Plant Biotechnol J 16:902–910
5. Wang W, Simmonds J, Pan Q et al (2018) Gene editing and mutagenesis reveal inter-cultivar differences and additivity in the contri-

bution of TaGW2 homoeologues to grain size and weight in wheat. Theor Appl Genet 131:2463–2475

6. Zhang Z, Hua L, Gupta A et al (2019) Development of an agrobacterium-delivered CRISPR/Cas9 system for wheat genome editing. Plant Biotechnol J 17:1623–1635
7. Wang Y, Cheng X, Shan Q et al (2014) Simultaneous editing of three homoeoalleles in hexaploid bread wheat confers heritable resistance to powdery mildew. Nat Biotechnol 32:947–951
8. Ishida Y, Tsunashima M, Hiei Y et al (2015) Wheat (Triticum aestivum L.) transformation using immature embryos. In: Wang K (ed) Agrobacterium protocols, vol vol. 1. Springer, New York, pp 189–198
9. Tian B, Navia-Urrutia M, Chen Y et al (2019) Biolistic transformation of wheat. Methods Mol Biol 1864:117–130
10. Tassy C, Barret P (2017) Biolistic transformation of wheat. Methods Mol Biol 1679: 141–152
11. Li J, Ye X, An B et al (2012) Genetic transformation of wheat: current status and future prospects. Plant Biotechnol Rep 6:183–193
12. Humphreys DG, Knox RE (2015) Doubled haploid breeding in cereals. In: Al-Khayri JM, Jain SM, Johnson DV (eds) Advances in plant breeding strategies: breeding, biotechnology and molecular tools. Springer International Publishing, Cham, pp 241–290
13. Ferrie AMR (2017) Doubled haploid production in higher plants. In: Thomas B, Murray BG, Murphy DJ (eds) Encyclopedia of applied plant sciences, vol 2. Academic Press, Waltham, MA, pp 147–151
14. Bhowmik P, Ellison E, Polley B et al (2018) Targeted mutagenesis in wheat microspores using CRISPR/Cas9. Sci Rep 8:6502
15. Doench JG, Fusi N, Sullender M et al (2016) Optimized sgRNA design to maximize activity and minimize off-target effects of CRISPR-Cas9. Nat Biotechnol 34:184–191
16. Doench JG, Hartenian E, Graham DB et al (2014) Rational design of highly active sgRNAs for CRISPR-Cas9-mediated gene inactivation. Nat Biotechnol 32:1262–1267
17. Rajagopalan N, Kagale S, Bhowmik P et al (2018) A two-step method for obtaining highly pure Cas9 nuclease for genome editing, biophysical, and structural studies. Methods Protoc 1:17
18. Sadasivaiah RS, Perkovic SM, Pearson DC, Postman B (2000) Registration of 'AC Nanda' wheat. Crop Sci 40:579–580
19. Pruitt SC, Mielnicki LM, Stewart CC (2004) Analysis of fluorescent protein expressing cells by flow cytometry. Methods Mol Biol 263:239–258
20. Hawley TS, Telford WG, Ramezani A et al (2001) Four-color flow cytometric detection of retrovirally expressed red, yellow, green, and cyan fluorescent proteins. BioTechniques 30:1028–1034
21. Thomas JB, Conner RL, Graf RJ (2012) Radiant hard red winter wheat. Can J Plant Sci 92:169–175

Chapter 16

Genetic Transformation of Protoplasts Isolated from Leaves of *Lolium temulentum* and *Lolium perenne*

Huw R. Davis, Anne L. Maddison, Dylan W. Phillips, and Huw D. Jones

Abstract

Transient expression of inserted recombinant DNA in plant protoplasts is a widely used tool for functional genomics research. Recently it has been utilized to screen potential sgRNA guides for CRISPR-mediated genome editing. However, little research has been conducted into the use of transient expression of protoplasts in *Lolium perenne* (a globally important pasture, hay, and amenity grass), and no studies have been conducted into *Lolium temulentum* (a weed in cereal crops but a potentially useful model species for *Lolium* research). In this chapter, we describe a methodology of protoplast extraction and transformation from 14-day-old leaf mesophyll cells from *L. perenne* and *L. temulentum*. We believe this is the first report of a procedure for obtaining high density, viable protoplasts from *L. temulentum*. The method of polyethylene glycol (PEG)-mediated transformation is also described to achieve genetic transformation of protoplasts.

Key words Protoplasts, *Lolium temulentum*, *Lolium perenne*, Transient expression, PEG-mediated transformation

1 Introduction

Transient gene expression of inserted recombinant DNA in protoplasts is a commonly used reverse genetic approach to study plant functional genomics [1–3]. In addition, over the last few years, it has been used to screen the efficiency of potential guide RNAs for CRISPR-mediated genome editing [4–6]. There are many different methods for achieving DNA transfer and transient expression in protoplasts however one of the most successful is PEG-mediated transformation which, at an alkaline pH, can be used to promote the binding and uptake of exogenous DNA to protoplasts.

In this method, it is thought that DNA adheres to the cell's surface and is absorbed by endocytosis or temporary disruption of the plasma membrane [1, 7–9]. The direct introduction of naked DNA into plant protoplasts commonly results in transient gene expression and can also, more rarely, result in integration of recombinant DNA

Luis M. Vaschetto (ed.), *Cereal Genomics: Methods and Protocols*, Methods in Molecular Biology, vol. 2072, https://doi.org/10.1007/978-1-4939-9865-4_16, © Springer Science+Business Media, LLC, part of Springer Nature 2020

into the host genome to create stably transformed cells. In some species these transformed protoplasts can be regenerated via specialized in vitro culture methods into whole plants.

The first reported experiment of plant protoplast isolation was recorded more than 50 years ago by Edward C. Cocking, and since then it has been used to observe cell division, embryogenesis, photosynthesis, and many more plant processes [8, 10]. There are many advantages of using PEG-mediated transformation, for example, it is easy to use, requires no specialist equipment, and is relatively inexpensive compared to other transformation methods [11]. PEG-mediated DNA delivery is particularly suitable as a rapid screen for transient expression but is less successful for routine, stable plant transformations.

Plant protoplasts are cells that have had their cell walls enzymatically removed, while they still retain the features and activities of other plant cells [12]. This feature has meant that protoplasts are excellent model systems to investigate cellular events such as transformation and recombinant DNA expression [12].

A recent scientific study reported the successful isolation and transformation of *L. perenne* protoplasts, which had previously not been achieved [13]. This study showed that protoplast transient expression can be used for functional gene analysis [13]. Previous studies had described the isolation of *L. perenne* protoplasts but with a relatively low success rate and with limited rates of transformation [13, 14].

To date, there are no publications describing the transformation of protoplasts isolated from *L. temulentum*, while studies have been conducted into developing transformation in *L. temulentum* using zygotic embryos [15]. This chapter describes the methods used to obtain a high yield of viable protoplasts and the methodology for high transformation efficiency for *L. temulentum*. The method described here is a protocol adapted from previously published studies. Broadly, the timings used for specific stages of the protoplast extraction steps followed Jung et al. [16]. While the composition of the various enzymes, buffers and methods of PEG-mediated transformation followed Yu et al. [13]. Here we report a successful methodology to extract and transform protoplasts from *L. perenne* and also apply it to successfully isolate and transform protoplasts of *L. temulentum* which has not previously been described.

The following method is the modified protocol for the preparation and extraction of protoplasts from 14-day old *L. temulentum* and *L. perenne* leaves. The chemical composition of solutions described in the Yu et al. [13] protocol achieved the greatest yield and viability, with the precise concentration of mannitol being a key factor for protoplasts viability. The timing of various mechanical operations and other methodological steps were adapted and optimized from Jung et al. [16].

Optimization over several replicate experiments demonstrated that this newly described protocol was able to achieve a high protoplast density and viability with a high transformation efficiency from both *L. temulentum* and *L. perenne.* An average yield of 4.88×10^7 and 6.18×10^7 per gram of leaf material and average viability of 74% and 77% were achieved *for L. perenne* and *L. temulentum* respectively using this newly described protocol. The following methodology achieved a transformation efficiency of 37% and 48% for *L. perenne* and *L. temulentum*. The fact that this protocol gave similarly good results for *L. perenne* and *L. temulentum*, leads us to believe that it would be a suitable starting point for the optimization of protoplast isolation and transformation in other *Lolium* species.

2 Materials

2.1 Plant Growth Conditions

Seeds of *L. perenne* (cv. AberDart) and *L. temulentum* (line BA3018) were grown in one part vermiculite, three parts perlite and nine parts peat moss in a controlled environment chamber, set at a constant of 23 °C with 20 h light (60–70% relative humidity). Plants were watered every 2 days and fertilized once a week with a ½ strength MS solution (*see* **Note 1**) [6].

2.2 Cell Wall Degrading Enzyme Solution

Leaf strip and protoplast extraction was achieved using the following solution: 10 mM MES, 20 mM KCl, 10 mM $CaCl_2$, mannitol (0.6 M), pH 5.7. Filter-sterilize the solution then store in 10 mL aliquots and defrost approximately 1 h before use. When ready to use add 1.5% cellulase, 0.75% macerozyme to the solution.

2.3 W5 Solution

W5 solution is used to release and wash protoplasts. To make the W5 solution prepare a solution of the following concentrations into 1 L of dH_2O: 2 mM MES, 154 mM NaCl, 125 mM $CaCl_2$, 5 mM KCL—correct the solution to pH 5.7. Filter-sterilize the solution and store at 4 °C

2.4 MMg Solution

MMg solution is used for the protoplast resuspension. 4 mM MES-KOH (pH 5.7), 0.4 M mannitol, 15 mM $MgCl_2$, pH 5.7. Filter-sterilize and store at 4 °C.

2.5 PEG-Ca^{2+} Solution

Used for the transformation of protoplasts. 20% (wt/vol) PEG4000, 100 mM $CaCl_2$, 300 mM mannitol. This was made in 50 mL quantities at one time; to 50 mL of dH_2O add 0.735 g $CaCl_2$, 2.732 g of mannitol, and 10 g of PEG400. Filter-sterilize and do not use for at least 1 h after preparation. Store at room temperature. It is recommended to make fresh on the day of use.

2.6 Fluorescein Diacetate (FDA) Stain

The FDA stain should be prepared immediately before use for protoplast viability. 20 μL of FDA stock in 1 mL 0.65 M mannitol. Then add equal volumes of FDA and isolated protoplast mix to a new Eppendorf. FDA is taken up by live cells.

3 Methods

3.1 Protoplast Isolation

1. Prepare 6 mL of cell wall degrading enzyme solution by adding 0.75% (wt/vol) macro enzyme and 1.5% (wt/vol) cellulose to the premade buffer, mix, and set aside.
2. Harvest 0.4 g from the second leaf of 14-day-old *L. perenne* (*see* Fig. 1) and/or *L. temulentum* plants. Using a sterile razor blade cut the leaves transversely into 0.5 mm sections then place in a sterilized 220 mL glass beaker containing 6 mL of filter-sterilized cell wall degrading enzyme solution.
3. Gently mix the leaves and enzyme solution by slowly swirling the mixture. Then vacuum infiltrate for 30 min at −15 mmHg pressure.
4. Incubate the sample at 30 °C in a nonshaking incubator for 60 min and then incubate in the dark, at room temperature for 18–20 h with gentle shaking at 35 rpm.
5. Presoak eight layers of autoclaved cheesecloth with chilled W5 solution and place in the top of a 50 mL tube. Gently pour the enzyme solution containing the leaves onto the cheesecloth to remove the larger particles. Wash the cheesecloth with 10 mL of chilled W5 (*see* **Note 2**) (*see* Fig. 1d).
6. Using a pasture pipette add an additional 5 mL of chilled W5 to the top of the enzyme solution, then leave the mixture in the dark, at room temperature for 1 h. This step allows the protoplasts to settle at the interface between the W5 and enzyme solution.
7. Collect the protoplasts using a pasture pipette and place into a new 50 mL tube containing 10 mL of chilled W5 solution.
8. Next, pellet the protoplasts by centrifugation for 5 min at 100 × *g* with slow-slow acceleration and deceleration speeds and remove the enzyme solution being careful to not disturb the protoplasts. Gently add 10 mL of chilled W5 to the tube repeating both the centrifuge and wash process once more.
9. After the second centrifuge step resuspend the protoplasts in 10 mL of W5 and leave at 4 °C in the dark overnight for the protoplasts to resuspend.
10. After 12 h or overnight remove the W5 solution and resuspend the protoplasts in 3 mL of chilled MMg solution.

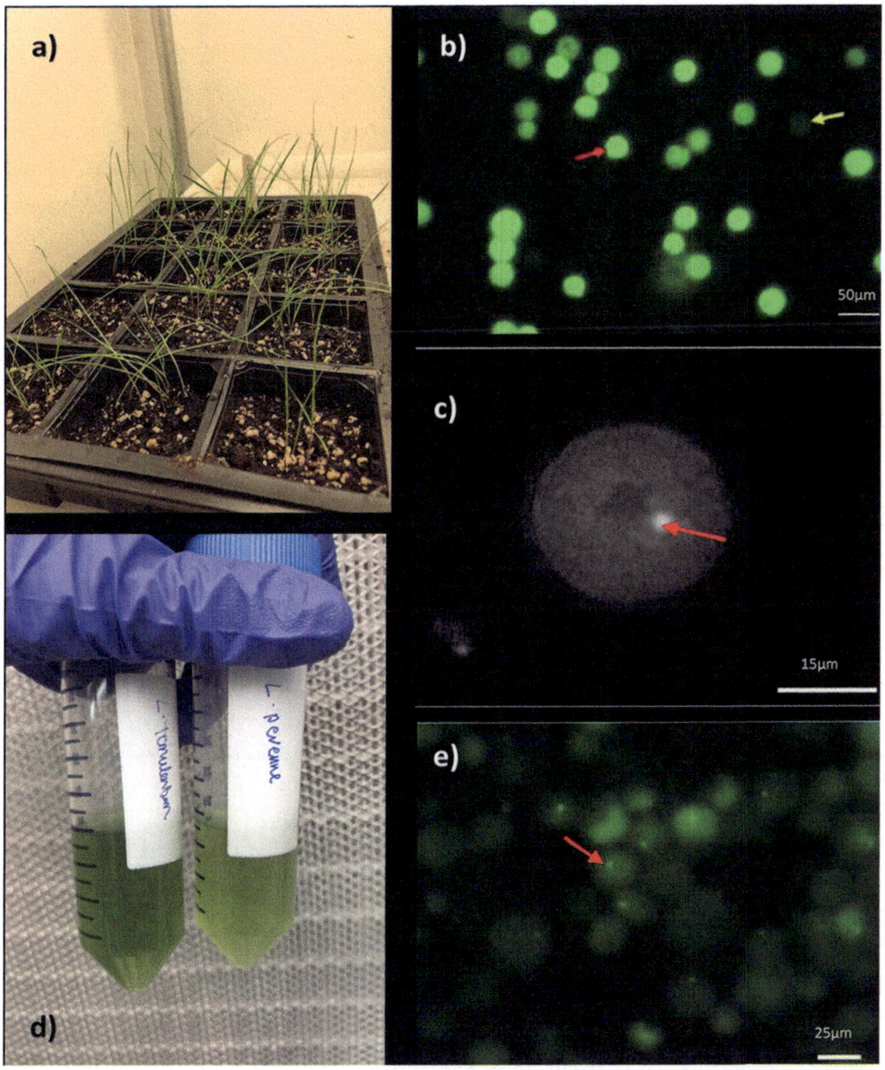

Fig. 1 Images from stages of the protoplast growth, isolation, and microscopy. (**a**) 14-day old ***L. perenne*** plants growing in a controlled environment chamber. (**b**) FDA viability stain of ***L. temulentum*** protoplasts. The red arrow shows a live FDA-stained protoplast; the yellow arrow shows a dead protoplast. (**c**) Transformed ***L. perenne*** protoplast, red arrow points to localization of GFP in the nucleus. (**d**) ***L. temulentum*** and ***L. perenne*** protoplast suspension. (**e**) Transformed ***L. temulentum*** protoplasts with GFP, red arrow shows the nuclear targeted GFP signal in the transformed nucleus

3.2 Protoplast Yield and Viability

Protoplast yield and viability were calculated by adding equal volumes of 1% FDA solution (*see* Subheading 2.6) and protoplast suspension (*see* Subheading 3.10). Add 20 μl of solution to a hemocytometer and examine with a fluorescent microscope. The FDA stain is up taken by live cells only (*see* Fig. 1b). Count the live and dead cells to determine the total yield as well as an estimated viability (*see* **Note 3**).

3.3 PEG-Mediated Transformation

1. Use 10 μL of 1 μg/μL plasmid DNA and add to 200 μL of protoplast stock (5 × 10^4) at room temperature and mix by flicking the tube.
2. Add 200 μL of PEG-Ca^{2+} (*see* Subheading 2.5) to protoplast and plasmid solution then mix by gently flicking the Eppendorf. Leave the solution in the dark, at room temperature for 10–15 min (*see* **Note 4**).
3. Add 3 mL of chilled W5 solution to the protoplast PEG/protoplast solution and centrifuge at 100 × *g* for 2 min to pellet the protoplasts.
4. Remove the supernatant and resuspend protoplasts in 0.5 mL of chilled W5 solution and leave for between 10 and 24 h in the dark, at room temperature in a 1.5 mL centrifuge tube (*see* **Note 5**).

3.4 Microscopy

Transformed protoplasts were viewed using a Leica DM6000B wide-field fluorescence microscope equipped with a Leica DFC350 FX R2 camera controlled by Leica LAS-AF software. Cellular localization of green florescent protein (GFP) can be observed in the nucleus of transformed protoplasts (*see* Fig. 1c, e).

4 Notes

1. To make the ½ strength MS solutions, add 2.15 g of Murashige and Skoog (MS) powder to 1 L of distilled H_2O and mix thoroughly.
2. It was also found that by gently squeezing the cheesecloth obtained a greater protoplast yield.
3. Useful online application that can help with counting and calculation yield/viability.
4. Below 10 min was insufficient time for the cells to uptake the plasmid, however over 15 min the PEG solution can cause cell death.
5. Studies suggest to precoat the 1.5 mL Eppendorfs with 1% BSA to avoid protoplast attachment; however, this was not found to be an issue. If protoplast attachment occurs, consider coating the tubes in BSA as research suggests.

Acknowledgments

The Institute of Biological, Environmental and Rural Sciences (IBERS) receives strategic funding from the Biotechnology and Biological Sciences Research Council (BBSRC) via grant [BB/CSP1730/1].

References

1. Cao Y, Li H, Pham AQ et al (2016) An improved transient expression system using *Arabidopsis* protoplasts. Curr Protoc Plant Biol 1:285–291
2. Zhang Y, Su J, Duan S et al (2011) A highly efficient rice green tissue protoplast system for transient gene expression and studying light/ chloroplast-related processes. Plant Methods 7:30
3. Sahab S, Hayden MJ, Mason J et al (2019) Mesophyll protoplasts and PEG-mediated transfections: transient assays and generation of stable transgenic canola plants. In: Transgenic plants. Humana Press, New York, pp 131–152
4. Zhang Y, Liang Z, Zong Y et al (2016) Efficient and transgene-free genome editing in wheat through transient expression of CRISPR/Cas9 DNA or RNA. Nat Commun 7:12617
5. Kim D, Alptekin B, Budak H (2018) CRISPR/ Cas9 genome editing in wheat. Funct Integr Genomics 18:31–41
6. Lin CS, Hsu CT, Yang LH et al (2018) Application of protoplast technology to CRISPR/Cas9 mutagenesis: from single-cell mutation detection to mutant plant regeneration. Plant Biotechnol J 16:1295–1310
7. Cao J, Yao D, Lin F et al (2014) PEG-mediated transient gene expression and silencing system in maize mesophyll protoplasts: a valuable tool for signal transduction study in maize. Acta Physiol Plant 36:1271–1281
8. Yoo S, Cho Y, Sheen J (2007) *Arabidopsis mesophyll* protoplasts: a versatile cell system for transient gene expression analysis. Nat Protoc 2:1565–1572
9. Dlugosz EM, Lenaghan SC, Stewart CN Jr (2016) A robotic platform for high-throughput protoplast isolation and transformation. J Vis Exp (115):54300. https://doi.org/10.3791/54300
10. Cocking EC (1960) A method for the isolation of plant protoplasts and vacuoles. Nature 187:962–963
11. Barampuram S, Zhang Z (2011) Recent advances in plant transformation. In: Birchler J (ed) Plant chromosome engineering, Methods in molecular biology (methods and protocols), vol 701. Humana Press, Totowa, pp 1–35
12. Wu F, Hanzawa Y (2018) A simple method for isolation of soybean protoplasts and application to transient gene expression analysis. J Vis Exp (131). https://doi.org/10.3791/57258
13. Yu G, Cheng Q, Xie Z et al (2017) An efficient protocol for perennial ryegrass mesophyll protoplast isolation and transformation, and its application on interaction study between LpNOL and LpNYC1. Plant Methods 13:46
14. Rathnam CK, Chollet R (1980) Photosynthetic and photorespiratory carbon metabolism in mesophyll protoplasts and chloroplasts isolated from isogenic diploid and tetraploid cultivars of ryegrass (Lolium perenne L.). Plant Physiol 65:489–494
15. Dalton SJ, Bettany AJE, Timms E et al (1999) Co-transformed, diploid *Lolium perenne* (perennial ryegrass), *Lolium multiflorum* (Italian ryegrass) and *Lolium temulentum* (darnel) plants produced by microprojectile bombardment. Plant Cell Rep 18:721–726
16. Jung HI, Yan J, Zhai Z et al (2015) Gene functional analysis using protoplast transient assays. Methods in molecular biology. Humana Press, New York, pp 433–452

Chapter 17

Agrobacterium Transformation in the Rice Genome

Asuka Nishimura

Abstract

Modification of the rice genome by *Agrobacterium*-mediated transformation is a general technique that can be easily performed today. Successful methods were established by vigorous studies on the culture system and the elucidation of *Agrobacterium* transformation mechanisms. This section provides a detailed description of routine and efficient rice transformation protocols by *Agrobacterium*. This method uses mature seeds as a material and can be applied to many *japonica* and some other varieties of rice. According to this method, it is even possible for beginners to obtain rice transformants.

Key words *Oryza sativa*, Japonica rice, *Agrobacterium*, Nuclear transformation, Embryogenic callus, Hygromycin selection

1 Introduction

Transformation methods are indispensable for the elucidation, modification, and editing of target genes and genomes. In rice, the basis of currently popular transformation methods mediated by *Agrobacterium tumefaciens* (syn. *Rhizobium radiobacter*) was developed about 25 years ago [1]. The important points for the success of this method are to use the scutellum-derived callus as the starting material and add an appropriate amount of acetosyringone at the time of infection. The transformation of *Agrobacterium* depends on the wounding responses of plants; the key processes are the production of virulence (*vir*) gene-inducing molecules and the rapid replication of introduced DNA. Monocotyledonous plants were originally outside the host range of *Agrobacterium*, so artificial induction was made possible by using a phenolic compound (acetosyringone) that is a trigger molecule for *vir* gene expression. As a result of testing the infection of various tissue pieces, the scutellum-derived embryogenic callus, comprising actively dividing cells, was identified as the optimum material for transformation in rice [1, 2]. However, as the production of the embryogenic callus is strongly restricted by genetic background [3–5], in the first test of rice transformation, it is recom-

Luis M. Vaschetto (ed.), *Cereal Genomics: Methods and Protocols*, Methods in Molecular Biology, vol. 2072,
https://doi.org/10.1007/978-1-4939-9865-4_17,

mended to use *japonica* rice varieties (e.g., Nipponbare, Taichung 65), which has been proven to be able to easily produce high-quality embryogenic calli derived from the scutella of mature seeds. As long as stable transformation using these rice varieties is successfully achieved, the transformation of other varieties will be possible by changing the start material, culture medium, and infection conditions [6–8]. This chapter describes a stable rice transformation protocol based on using standard *japonica* rice (*see* **Note 1**). For the outline of the protocol, it is first essential to grow healthy rice plants to obtain good seeds, and prepare the culture conditions. Obtaining excellent-quality embryonic calli is important for success. Then, infection treatment is carried out using a small amount of *Agrobacterium* in the presence of acetosyringone, using actively cell-dividing calli derived from seeds as a material. After infection treatment, transformants can be produced by culturing for an appropriate amount of time using an appropriate medium.

2 Materials

2.1 Plant Material and Agrobacterium tumefaciens

1. Plant Material. Mature dry seeds of *japonica* rice, Nipponbare (*see* **Note 1**): in warmer seasons suitable for cultivation, grow rice healthily in the field or a greenhouse, and harvest high-quality mature seeds without damage. After harvesting, place the seeds in a paper bag and dry at 20–25 °C for 7–10 days (using a desiccator at room temperature for convenience) (*see* **Note 2**). Store dried seeds in a sealed container together with silica gel in a refrigerator to avoid reabsorption of moisture (*see* **Note 3**).
2. *Agrobacterium tumefaciens*. Prepare competent cells of *Agrobacterium* EHA105 [9] by conventional methods [10] (*see* **Note 4**). Transform the binary vector containing the hygromycin resistance gene (*hpt*) driven by the CaMV 35S promoter for plant selection and the kanamycin resistance gene (*nptII*) for bacterial selection (*see* **Note 5**) into the *Agrobacterium* using electroporation or the freeze–thaw method [10].

2.2 Stock Solutions for Plant Culture Media

2.2.1 Vitamins and Phytohormones

1. N6-Vitamins: dissolve 100 mg of glycine, 25 mg of nicotinic acid, 25 mg of pyridoxine hydrochloride, and 50 mg of thiamine hydrochloride in ultrapure water, make up to 500 mL with the water, and store at 4 °C.
2. MS-Vitamins: dissolve 250 mg of nicotinic acid, 500 mg of pyridoxine hydrochloride, 500 mg of thiamine hydrochloride, and 100 mg of glycine in ultrapure water, make up to 500 mL with the water, and store at 4 °C.
3. 2,4-Dichlorophenoxyacetic acid (2, 4-D; 0.2 mg/mL solution): dissolve 100 mg of 2,4-D in 1 mL of 1 N NaOH, make up to 500 mL with ultrapure water, and store at 4 °C.

4. 1-Naphthylacetic acid (NAA; 0.2 mg/mL solution): dissolve 100 mg of NAA in 1 mL of 1 N NaOH, make up to 500 mL with ultrapure water, and store at 4 °C.
5. Kinetin, 0.2 mg/mL solution: dissolve 100 mg of kinetin in 20 mL of 0.1 N HCl, make up to 500 mL with ultrapure water, and store at 4 °C.

2.2.2 Antibiotics

1. 50 mg/mL meropenem solution: dissolve in 0.1 M Na_2CO_3, filter-sterilize, and store at −20 °C.
2. 50 mg/mL hygromycin solution: dissolve hygromycin B in ultrapure water, filter-sterilize, and store at −20 °C.

2.3 Media for Plant Culture (Modified Based on [11])

All solid media should be prepared before use and stored at room temperature (20–25 °C) when not in use immediately (*see* **Note 6**).

1. Callus induction medium: mix and dissolve 4.1 g of CHU (N6) basal salt mixture (FUJIFILM Wako Pure Chemical, Osaka, Japan [12]), 10 mL of N6-Vitamins, 0.1 g of myo-inositol, 0.3 g of casamino acids, 2.878 g of proline, 10 mL of 2,4-D, and 30 g of sucrose in 800 mL of ultrapure water. Adjust the pH to 5.8, add 3 g of gellan gum, make up to 1 L with ultrapure water, and autoclave. Cool the medium to about 50 °C and pour into 25 sterile plastic petri plates (90 × 20 mm) in a laminar flow hood, solidify, and dry (*see* **Note 7**).
2. Selection medium: make the callus induction medium and add 1 mL of meropenem and 1 mL of hygromycin before pouring (*see* **Note 7**).
3. Regeneration medium: mix and dissolve 4.6 g of MS plant salt mixture (FUJIFILM Wako Pure Chemical, Osaka, Japan [13]) 10 mL of MS-Vitamins, 0.1 g of myoinositol, 2 g of casamino acid, 1 mL of NAA, 10 mL of kinetin, 30 g of sucrose, and 30 g of sorbitol in 800 mL of ultrapure water. Adjust the pH to 5.8, add 3 g of gellan gum, make up to 1 L with ultrapure water, and autoclave. Cool the medium to about 50 °C and add 0.5 mL of meropenem and 1 mL of hygromycin, pour into 25 sterile plastic petri plates (90 × 20 mm) in a laminar flow hood, solidify and dry (*see* **Note 7**).
4. Rooting medium: Mix and dissolve 4.6 g of MS plant salt mixture, 10 mL of MS-Vitamins, and 30 g of sucrose in 800 mL of ultrapure water. Adjust the pH to 5.8, add 2.6 g of gellan gum, make up to 1 L with ultrapure water, and autoclave. Cool the medium to about 50 °C, add 1 mL of hygromycin, pour into sterilized plant culture boxes in a laminar flow hood, solidify, and dry.

2.4 Stock Solutions for Agrobacterium Infection

1. 15 mg/mL rifampicin solution: dissolve rifampicin in dimethyl sulfoxide (DMSO), filter-sterilize, and store at −20 °C.
2. 50 mg/mL kanamycin solution: dissolve kanamycin sulfate with ultrapure water, filter-sterilize, and store at −20 °C.
3. AA solutions (*see* Table 1): prepare all stock solutions using ultrapure water and store at 4 °C.
4. 100 mg/mL L-cysteine solution (make before use): dissolve L-cysteine in ultrapure water and filter-sterilize.
5. 10 mg/mL acetosyringone solution: dissolve with DMSO, filter-sterilize, and store at −20 °C.

2.5 Media for Agrobacterium Infection

1. Lysogeny broth (LB) medium: mix and dissolve 1 g of tryptone, 0.5 g of yeast extract, 1 g of NaCl, and 20 μL of 5 N NaOH in 90 mL of ultrapure water. Add 1.5 g of agar, make up to 100 mL with ultrapure water, and autoclave. Cool the medium to about 50 °C, add 0.1 mL of rifampicin and 0.1 mL kanamycin (*see* **Note 5**), pour into sterile plastic petri plates (90 × 15 mm) in a laminar flow hood, solidify, and dry.
2. Infection medium (liquid) [1]: mix and dissolve the following reagents in 400 mL of ultrapure water: 0.5 mL each of AA-1–5, 2.5 mL of AA-6, 5 mL of AA-Sol, 0.25 g of casamino acids, 34.25 g of sucrose, 18 g of glucose, 0.45 g of L-glutamine, 0.15 g of L-aspartic acid, and 1.5 g of potassium chloride. Adjust the pH to 5.2, make up to 500 mL with ultrapure water, autoclave, and store at 4 °C.
3. Cocultivation medium (liquid) [14]: mix the same reagents as those in the callus induction medium (except for gellan gum) in 400 mL of ultrapure water and add 2.5 g of glucose. Adjust the pH to 5.8, make up to 500 mL with ultrapure water, autoclave, and store at 4 °C.

2.6 Other Reagents and Laboratory Supplies

1. Ethanol: 70% (vol/vol) in distilled water.
2. 1.5% sodium hypochlorite in sterile distilled water, prepared before use.
3. Small rice husker.
4. Surgical tape.
5. Parafilm.
6. Sterile plastic petri plates: 90 × 20 mm.
7. Screw cap plastic tubes: 50 mL, transparent.
8. Microspatula or loop.
9. Stainless steel sieve (or tea strainer).
10. Paper towels.
11. Filter paper: 90 mm in diameter, No. 2.

Table 1
AA stock solutions

Solution	Components	Amount (per 100 mL)
AA-1	Manganese (II) sulfate pentahydrate	1000 mg
	Boric acid	300
	Zinc sulfate heptahydrate	200
	Disodium molybdate (VI) dihydrate	25
	Copper (II) sulfate pentahydrate	2.5
	Cobalt (II) chloride hexahydrate	2.5
	Potassium iodide	75
AA-2	Calcium chloride dihydrate	15 g
AA-3	Magnesium sulfate heptahydrate	25 g
AA-4	Ethylenediamine-N,N,N′,N′-tetraacetic acid, Iron (III) sodium salt, trihydrate	4 g
AA-5	Sodium dihydrogen phosphate dihydrate	15 g
AA-6	Nicotinic acid	20 mg
	Pyridoxine hydrochloride	20
	Thiamine hydrochloride	200
	Myo-inositol	2000
AA-Sol	L-arginine	1.8 g
	Glycine	0.075

3 Methods

3.1 Callus Induction from Mature Rice Seeds

1. Remove the hulls from the required amount of mature rice seeds and collect healthy seeds with intact embryos in a 50 mL tube (*see* **Note 8**).
2. Add 70% ethanol to the tube, hand-shake for 30 s to sterilize the surface, and discard the liquid.
3. Pour 1.5% sodium hypochlorite to below half of the tube and vigorously shake for 20–30 min.
4. In the laminar flow hood, discard the sodium hypochlorite solution and wash the seeds five times with sterilized ultrapure water.
5. In the laminar flow hood, place 10–15 sterile seeds at intervals with the embryo pointing upward, slightly embedded in the callus induction medium (*see* **Note 9**). Seal the plate with surgical tape.
6. Incubate the plates under dim light (about 3.5 klux) at 28–30 °C for 3–4 weeks (*see* **Note 10**).

3.2 Preculture of *Agrobacterium* and Calli for Infection

1. Streak a single colony of transformed *Agrobacterium* onto the LB medium. Incubate in darkness at 28 °C for 3 days.
2. In the laminar flow hood, collect the calli spilling from the scutellum for three plates (*see* Subheading 3.1) onto one fresh callus induction medium (yellowish white calli, slightly transparent, about 1.5–3 mm in diameter), and incubate under dark or dim light (about 3.5 klux) at 28–30 °C for 3 days (*see* **Note 11**).

3.3 Callus Infection by *Agrobacterium* (Modified Based on [14])

1. Add 30 mL of infection medium and 45 μL of acetosyringone to a 50 mL tube.
2. Scrape a small amount of precultured *Agrobacterium* with a microspatula or loop and sufficiently suspended in the infection medium so that aggregates do not remain (*see* **Note 12**). Adjust the *Agrobacterium* optical density (OD600) to 0.05–0.1.
3. Prepare 11 mL of cocultivation medium in a 50 mL tube, add 16.5 μL of acetosyringone and 11 μL of fresh L-cysteine. Pour 5.5 mL of the prepared cocultivation medium into two sterilized petri plates containing three sterilized filters. Seal with Parafilm until use (*see* **Note 13**).
4. Collect the calli on the preculture plate in one place and place in a new 50 mL sterile tube. Pour the whole amount of the infection medium (prepared in **steps 1** and **2** above) into the tube and shake gently for 2 min.
5. Place a stainless-steel sieve above an empty sterile petri plate during the waiting time.
6. Pour the calli over the sieve, lift the calli with the sieve, place them on sterile paper towels and remove the *Agrobacterium* fluid.
7. Place calli onto two cocultivation plates (prepared in **step 3** above) and seal with Parafilm.
8. Incubate in darkness at 25 °C for 3 days.

3.4 Selection of Transformed Calli

The following **steps 1–4** should be performed in a laminar flow hood.

1. Place the infected calli (on the two coculture plates) into a 50 mL sterile transparent tube and wash with sterile water by gently shaking the tube. Discard the suspension of *Agrobacterium* slowly into a beaker. Repeat washes until the rinse water is clear (*see* **Note 14**).
2. Rinse the calli with sterile water to which meropenem was added at a 1000-fold dilution. Pour the calli suspension into a sterile stainless-steel sieve placed on a plastic petri dish.

3. Put the stainless-steel sieve containing calli onto sterile paper towels and remove water from the surface of the calli (*see* **Note 15**).
4. Transfer about 20 calli onto a selection media (*see* **Note 16**) and seal with surgical tape.
5. Incubate at 28–30 °C under dark or about 3.5 klux for 3–4 weeks until the selected calli grow (*see* **Notes 17** and **18**).

3.5 Regeneration of Transgenic Rice Plants

The following **steps 1** and **2** should be performed in a laminar flow hood.

1. Transfer the growing calli to regeneration medium, place four or five calli populations originating from a single cocultured callus on one plate (*see* **Note 19**).
2. Seal plates with surgical tape.
3. Incubate at 28–30 °C and about 4.5 klux for 3–4 weeks.

3.6 Rooting of Transgenic Rice Plants

The following **steps 1** and **2** should be performed in a laminar flow hood.

1. Transfer regenerated shoots of 3–4 cm in length to rooting medium in plant boxes (*see* **Note 20**).
2. Seal boxes with surgical tape, if necessary.
3. Incubate at 28–30 °C and about 4.5 klux for several weeks until the roots grow sufficiently.

3.7 Habitation

1. When the plant grows sufficiently in the boxes, open the lid slightly and gradually expose it to the outside air.
2. After confirming the growth of the plant, gently pull out the plant from the gel medium and thoroughly remove the gel adhering to the roots.
3. If necessary, cultivate with soil and water.

4 Notes

1. It is possible to use *japonica* varieties other than Nipponbare, but it is preferable to use a variety that efficiently obtains regenerated plants in the antibiotic-free medium used in this protocol. For Koshihikari and its related strains, which are elite varieties in Japan, it is necessary to modify the culture medium, as the nitrogen metabolizing ability is lowered [3]. For these cultivars, this method can be used by using a low-nitrogen medium. Specifically, in the case of N6, 707.5 mg/L of potassium nitrate and 688 mg/L of L-alanine are added to a nitrogen-free N6, and a medium supplemented with 475 mg/L of potassium nitrate and 412.5 mg/L of ammonium nitrate is used for a nitrogen-free MS.

2. When dried at high temperatures, the germination rate decrease quickly and should be dried slowly at a low temperature.
3. The germination rate of dried seeds is maintained for more than 10 years by storage at −1 to 5 °C and 50% relative humidity (germinability depends on variety).
4. Generally, a binary vector system is used, in which a helper plasmid having a series of gene introductions for the *vir* gene and a plasmid having the gene to be introduced in the T-DNA region are separate. In addition to the EHA 105 strain, EHA 101 and LBA 4404 [15] are also frequently used.
5. For bacteria selection, other antibiotics can be used, but hygromycin can be selected most efficiently for rice. As various vector systems have been developed, appropriate reagents should be selected according to the vector and *Agrobacterium* strain used.
6. In the case of rice, lowering the water content of the callus is considered to be important for successful culturing. It is advisable to keep the petri medium sufficiently dry and store with the humidity lowered.
7. Since sterile plastic petri dishes are needed for rice tissue culture, there is concern about the effect of residual ethylene oxide gas (EOG) on the culture; thus, we recommend using gamma ray-sterilized petri dishes instead of EOG-sterilized dishes.
8. Usually 30–50 seeds per infection are used.
9. By placing each seed at a distance from each other, it is possible to rescue the remaining seeds from fungus-contaminated seeds during culturing.
10. Observe the state of the plate frequently about 1 week after the start of culturing and check for contamination. When contamination occurs, the remaining uncontaminated seeds are immediately transplanted into a fresh medium.
11. Before collecting the calli for subculture, remove the seeds that have developed seedlings or brown calli from the culture plates and discard them.
12. If *Agrobacterium* aggregates remain, excessive proliferation occurs during coculture and the infected calli are damaged.
13. The antioxidant effect of L-cysteine is thought to reduce plant cell damage caused by *Agrobacterium* infection, and if it is converted to cystine by air oxidation its effect will be lost.
14. Usually, it becomes a clear solution by washing five or six times. For subsequent easy handling, wash gently so that the callus does not break.

15. It is important to remove excess water from the calli to ensure active culture growth.
16. In the process of selective culture, as the callus proliferation progresses, adjacent calli can mix. Place each callus piece apart so that independent transformation events can be distinguished.
17. *Agrobacterium* may grow on the callus surface during the beginning of selective culture. Observe the calli every day; if *Agrobacterium* proliferates, transfer only other healthy calli to the new selection medium. Additionally, as the selection culture period becomes longer, the effect of the antibiotics weakens, and *Agrobacterium* regrows. If the transformed callus proliferates slowly, it is recommended to transfer to a new selection medium.
18. Nontransformed cells are browning and do not proliferate, but most calli proliferate when the transformation efficiency is good.
19. If calli are placed on the regeneration medium at high density, the regeneration efficiency will decline. Pick about five healthy pieces from each growing calli population and place them on the regeneration medium to distinguish each population.
20. Often, not only shoots but also roots are generated during regeneration culture. Regenerated shoots with roots are transplanted to the rooting medium, subsequent growth is smooth.

Acknowledgments

The author would like to thank Dr. Makoto Matsuoka, Dr. Nobuhiro Tsutsumi, and the Honda Research Institute and Japan Society for Promotion of Science (JSPS) Fellowship for supporting research on this protocol.

References

1. Hiei Y, Ohta S, Komari T et al (1994) Efficient transformation of rice (*Oryza sativa* L.) mediated by Agrobacterium and sequence analysis of the boundaries of the T-DNA. Plant J 6:271–282
2. Hiei Y, Komari T, Kubo T (1997) Transformation of rice mediated by Agrobacterium tumefaciens. Plant Mol Biol 35: 205–218
3. Nishimura A, Ashikari M, Lin S et al (2005) Isolation of a rice regeneration quantitative trait loci gene and its application to transformation systems. Proc Natl Acad Sci USA 102:11940–11944
4. Ge X, Chu Z, Lin Y et al (2006) A tissue culture system for different germplasms of indica rice. Plant Cell Rep 25:392–402
5. Zhao L, Zhou H, Lu L et al (2009) Identification of quantitative trait loci controlling rice mature seed culturability using chromosomal segment substitution lines. Plant Cell Rep 28:247–256
6. Rashid H, Yokoi S, Toriyama K et al (1996) Transgenic plant production mediated by Agrobacterium in indica rice. Plant Cell Rep 15:727–730
7. Lin YJ, Zhang QF (2005) Optimising the tissue culture conditions for high efficiency trans-

formation of indica rice. Plant Cell Rep 23:540–547

8. Hiei Y, Komari T (2006) Improved protocols for transformation of indica rice mediated by Agrobacterium tumefaciens. Plant Cell Tissue Organ Cult 85:271–283
9. Hood EE, Gelvin SB, Melches LS et al (1993) New Agrobacterium helper plasmids for gene transfer to plants. Transgenic Res 2:208–218
10. Wise AA, Liu Z, Binns AN (2006) Three methods for the introduction of foreign DNA into Agrobacterium. In: Wang K (ed) Agrobacterium protocols. Methods in molecular biology, vol 343. Humana Press, New York, pp 43–53
11. Toki S (1997) Rapid and efficient agrobacterium-mediated transformation in rice. Plant Mol Biol Rep 15:16–21
12. Chu CC, Wang CC, Sun CS (1975) Establishment of an efficient medium for anther culture of rice through comparative experiments on the nitrogen sources. Sci Sinica 18:659–668
13. Murashige T, Skoog F (1962) A revised medium for rapid growth and bioassays with tobacco tissue cultures. Physiol Plant 15: 473–497
14. Ozawa K (2009) Establishment of a high efficiency Agrobacterium-mediated transformation system of rice (*Oryza sativa* L.). Plant Sci 176:522–527
15. Hood EE, Helmer GL, Fraley RT et al (1986) The hypervirulence of Agrobacterium tumefaciens A281 is encoded in a region of pTiBo542 outside of T-DNA. J Bacteriol 168: 1291–1301

Chapter 18

In Vivo Phosphorylation: Development of Specific Antibodies to Detect the Phosphorylated PEPC Isoform for the C4 Photosynthesis in *Zea mays*

Yoshihisa Ueno, Kumiko Yoshizawa-Kumagaye, Junji Emura, Tomoko Urabe, Taku Yoshiya, Tsuyoshi Furumoto, and Katsura Izui

Abstract

Phosphoenolpyruvate carboxylases (PEPCs), mostly known as the enzymes responsible for the initial CO_2 fixation during C_4 photosynthesis, are regulated by reversible phosphorylation in vascular plants. The phosphorylation site on a PEPC molecule is conserved not only among isoforms but also across plant species. An anti-phosphopeptide antibody is a common and powerful tool for detecting phosphorylated target proteins with high specificity. We generated two antibodies, one against a peptide containing a phosphoserine (phosphopeptide) and the other against a peptide containing a phosphoserine mimetic, (*S*)-2-amino-4-phosphonobutyric acid (phosphonopeptide). The amino acid sequence of the peptide was taken from the site around the phosphorylation site near the *N*-terminal region of the maize C_4-isoform of PEPC. The former antibodies detected almost specifically the phosphorylated C_4-isoform of PEPC, whereas the latter antibodies had a broader specificity for the phosphorylated PEPC in various plant species. The following procedures are described herein: (1) preparation of the phosphopeptide and phosphonopeptide; (2) preparation and purification of rabbit antibodies; (3) preparation of cell extracts from leaves for analyses of PEPC phosphorylation with antibodies; and (4) characterization of the obtained antibodies. Finally, (5) two cases involving the application of these antibodies are presented.

Key words PEPC, Phosphoenolpyruvate carboxylase, C_4 photosynthesis, Protein phosphorylation, Phosphopeptide antibody, Phosphonopeptide antibody, Synthetic peptide, Immunodetection, *Zea mays*, *Flaveria bidentis*

1 Introduction

Phosphoenolpyruvate carboxylase (PEPC; EC 4.1.1.31), which is usually composed of four identical subunits with an *Mr* of approximately 100 kDa, is a CO_2-fixing enzyme that produces oxaloacetate from phosphoenolpyruvate (PEP) and bicarbonate, liberating Pi in the presence of Mg^{2+} (*for reviews see* [1–3]). Phosphoenolpyruvate carboxylase is widespread not only in all photosynthetic organisms, such as plants, algae, cyanobacteria,

Luis M. Vaschetto (ed.), *Cereal Genomics: Methods and Protocols*, Methods in Molecular Biology, vol. 2072, https://doi.org/10.1007/978-1-4939-9865-4_18,

and photosynthetic bacteria, but also in most nonphotosynthetic bacteria and protozoa; however, it is apparently absent in animals, yeasts, and fungi. This enzyme primarily plays an anaplerotic role by replenishing C_4-dicarboxylic acids used for both energy and biosynthetic metabolism. Additionally, PEPC has multiple physiological roles in plants. For example, it is responsible for the initial CO_2-fixation in the photosynthesis of C_4 and Crassulacean acid metabolism plants, represents a carbon source for symbiotic N_2-fixing bacteria in the nodule, and mediates the accumulation of malate and/or citrate in fruits [1–3]. These roles are shared by several isoforms with diverse catalytic and regulatory properties. Most PEPCs are regulated in an allosteric manner with various effectors depending on the species. The PEPCs of vascular plants are reversibly activated by the phosphorylation of the conserved Ser located near the *N*-terminus [1, 2, 4]. The phosphorylation is catalyzed by a specific protein kinase, PEPC-PK [5–7], whereas the dephosphorylation of the enzyme is mediated by protein phosphatase 2A, with a moderate substrate specificity [8]. The structural mechanism underlying the phosphorylation-induced enzymatic activation remains unclear [1, 2, 4, 9, 10].

To elucidate the dynamics of the phosphorylation of a target protein in vivo in response to changes in the internal or external environmental stimuli, several methods have been developed that are specific enough to detect the phosphorylated target proteins among many other phosphoproteins. An immunoblotting technique seems to be widely used for this purpose. In this method, antibodies specific for the phosphorylated site of target proteins are required. The antibodies are raised against synthetic phosphorylated oligopeptides with the amino acid sequence surrounding the phosphorylation site. Because there is usually a uniquely conserved amino acid sequence around the phosphorylation site of the target protein, the antibodies raised against the target protein of one species may exhibit moderately high and specific affinity for the corresponding proteins of other species. We previously produced an antibody specific for the *N*-terminal Ser (Ser^{15})-phosphorylated C_4-form of PEPC in maize [11]. The antibody was raised against a peptide containing phosphoserine (phosphopeptide). Despite the limited titer, the specificity of this antibody was satisfactory. This phosphopeptide antibody is hereafter abbreviated as the PC4 antibody. In transgenic rice plants carrying the maize PEPC gene, the phosphorylated maize C_4-form of PEPC was specifically detected with this antibody without the undesired detection of the phosphorylated PEPCs encoded by endogenous rice genes [12–14], confirming the amino acid sequence-specificity of this antibody.

We raised another antibody against a peptide containing a phosphoserine mimetic, (*S*)-2-amino-4-phosphonobutyric acid (L-Apb) (phosphonopeptide), to avoid the dephosphorylation of antigens by endogenous phosphatases during the immunization of rabbits. The obtained antibody had a higher affinity than the PC4 antibody and recognized the phosphorylated PEPCs in maize and in various vascular plants [15–18]. This phosphonopeptide antibody is hereafter referred to as the PNB antibody.

We herein describe improved protocols for the syntheses of a phosphopeptide and a phosphonopeptide as an alternative antigen. We also describe the protocol for raising antibodies against these peptides as well as our protocols for the preparation of cell extracts and the immunodetection of phosphorylated PEPC with the antibodies. Moreover, we characterize and evaluate the utility of the antibodies. Finally, we present the following two cases in which these antibodies were applied: (1) an analysis of the dynamics of diurnal changes in the phosphorylation state of the maize C_4-form of PEPC with the PC4 antibody and (2) an examination of transgenic Flaveria plants lacking phosphorylated PEPC because of the knockdown of a gene encoding a specific protein kinase, PEPC-PK, with the PNB antibody.

2 Materials

2.1 Preparation of Antibodies Detecting Phosphorylated PEPC

2.1.1 Materials for the Synthesis of the Phosphopeptide

1. Automated peptide synthesizer.
2. Resins: *N*-9-Fluorenylmethoxycarbonyl (Fmoc) amino acid-preloaded 4-(hydroxymethyl)phenoxymethyl resin (Wang resin) or 4-(2′,4′-dimethoxyphenyl-*N*-Fmoc-aminomethyl) phenoxy resin (Rink amide resin).
3. Fmoc amino acids: Side chain protections are as follows: Ser[PO(OH)OBzl], Thr[PO(OH)OBzl], Tyr[PO(OH)OBzl], Asp(O*t*Bu), Glu(O*t*Bu), Arg(Pbf), Lys(Boc), His(Trt), Cys(Trt), Ser(*t*Bu), Thr(*t*Bu), Tyr(*t*Bu), Trp(Boc), Asn(Trt), Gln(Trt).

   Abbreviations: OBzl, O-benzyl; O*t*Bu, O-*tert*-Butyl; Pbf, 2,2,4,6,7-pentamethyl-dihydrobenzofuran-5-sulfonyl; Boc, *tert*-butoxycarbonyl; Trt, trityl.
4. Preparative column and instrument for reversed-phase (RP) HPLC
5. Reagents and solvents:
   - Piperidine.
   - *N*-Methylpyrrolidone (NMP).
   - *O*-(6-Chlorobenzotriazol-1-yl)-*N*,*N*,*N*′,*N*′-tetramethyluronium hexafluorophosphate (HCTU).
   - 6-Chloro-1-hydroxybenzotriazole (6-Cl-HOBt).
   - *N*,*N*-Diisopropylethylamine (DIEA).

- Acetic anhydride (Ac_2O).
- 1-Hydroxybenzotriazole (HOBt).
- Dichloromethane.
- Diethyl ether.
- Trifluoroacetic acid (TFA).
- Triisopropylsilane (TIS).
- Thiophenol.
- Acetonitrile (CH_3CN).

2.1.2 Materials for the Conjugation of the Phosphopeptide Antigen to the Carrier Protein

1. Keyhole limpet hemocyanin (KLH, Calbiochem).
2. Phosphate-buffered saline (PBS) and PBS (pH 8.0). The pH of PBS is 7.0 to 7.4 unless otherwise stated.
3. *N*-(6-Maleimidocaproyloxy)succinimide (EMCS).
4. Dimethyl sulfoxide (DMSO).
5. 8 M urea (pH 7).
6. Dialysis tube.

2.1.3 Materials for Immunization

1. Japanese white rabbits.
2. Freund's adjuvant.
3. Cooling centrifuge.

2.1.4 Materials for the Precipitation of Antibodies with Ammonium Sulfate

1. Saturated ammonium sulfate solution.
2. PBS.
3. Dialysis tube.
4. Cooling centrifuge.

2.1.5 Materials for the Preparation of Affinity Columns

1. Amino-terminal cross-linked agarose gel (Affi-Gel 102 Gel, Bio-Rad).
2. PBS and PBS (pH 8.0).
3. EMCS.
4. DMSO.
5. Blocking buffer (0.1 M mercaptoethylamine in PBS).
6. End-over-end shaker.
7. Empty column.

2.1.6 Materials for the Affinity Purification of Antibodies

1. 1 M NaCl in phosphate buffer (pH 7).
2. PBS.
3. 0.1 M glycine–HCl (pH 2.3).
4. 0.5 M Tris–HCl (pH 7.5).

2.1.7 Materials for Enzyme-Linked Immunosorbent Assay (ELISA)

1. Ovalbumin (OVA).
2. PBS and PBS (pH 8.0).
3. EMCS or *N*-(4-maleimidobutyryloxy)succinimide (GMBS).
4. DMSO.
5. 8 M urea (pH 7).
6. Dialysis tube.
7. 96-well microtiter plate.
8. Coating buffer (carbonate buffer, pH 9.5).
9. Blocking solution (casein in PBS).
10. Horseradish peroxidase-conjugated secondary antibody.
11. PBS containing bovine serum albumin (BSA).
12. 2,2′-Azino-bis(3-ethylbenzothiazoline-6-sulfonic acid) diammonium salt (ABTS).
13. 50 mM acetate buffer (pH 4.0).
14. Plate reader.

2.1.8 Materials for the Production of a Phosphorylation Site-Specific Antibody Using the Phosphonopeptide as an Antigen

1. Automated peptide synthesizer (applicable to Boc chemistry).
2. Hydrogen fluoride (HF)-reaction apparatus.
3. Reversed-phase (RP) HPLC.
4. Boc-Arg(Tos)-4-oxymethylphenylacetamido-methyl (PAM) resin.
5. Boc-L-Apb(Me_2) (Fig. 1) [21].
6. Boc amino acids: Side chain protections are as follows: Asp(OHex), Glu(OHex), Arg(Tos), Lys(ClZ), His(Bom), Cys(MeBzl), Ser(Bzl), Thr(Bzl), Tyr(BrZ), Trp(For).

   Abbreviations: OHex; Ocyclohexyl, Tos; tosyl, ClZ; 2-chlorobenzyloxycarbonyl, Bom; benzyloxymethyl, MeBzl; 4-methylbenzyl, BrZ; 2-bromobenzyloxycarbonyl, For; formyl
7. Reagents and solvents:
   - 50% TFA/CH_2Cl_2.
   - DIEA.
   - DCC (*N*,*N*′-dicyclohexylcarbodiimide).
   - HOBt.
   - NMP.
   - Ac_2O.
   - 1-Ethyl-3-(3-dimethylaminopropyl)carbodiimide (WSCD).
   - DMF.
   - HF.

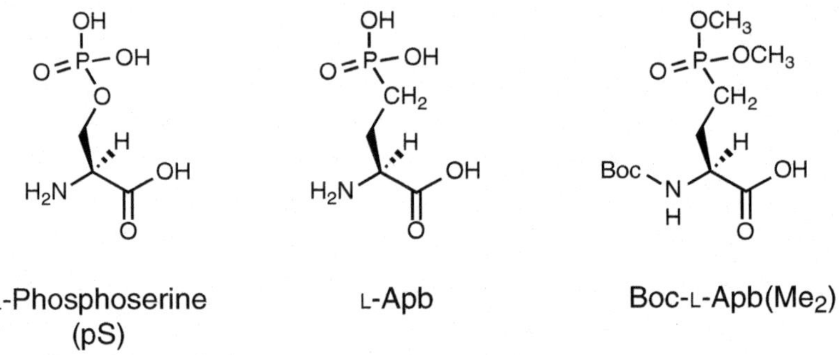

Fig. 1 Chemical structures of L-phosphoserine (pS), (*S*)-2-amino-4-phosphonobutyric acid (L-Apb), and Boc-L-Apb(Me_2)

- *p*-Cresol.
- Diethyl ether.
- Dithiothreitol.
- Aqueous ammonia.
- TFA.
- CH_3CN.
- Trimethylsilyl trifluoromethanesulfonate (TMSOTf).
- Dimethyl sulfide.
- Diisopropyl ether.

2.2 Detection of Phosphorylated PEPCs Using the Antibodies

2.2.1 Plant Materials

Maize: *Zea mays* inbred line H84 (*see* **Note 1**).

2.2.2 Materials for the Extraction of Proteins from Maize Leaves

Materials for the large-scale and microscale protein extraction methods are listed (*see* **Note 2**).

2.2.3 Large-Scale Extraction

1. Liquid nitrogen.
2. Cold room (9 °C or lower).
3. (Optional) Kitchen knife and a Waring blender.
4. Extraction buffer for a large-scale preparation (*see* **Note 3**).
   0.1 M Tris–HCl, pH 7.5.
   10 mM EDTA (*see* **Note 4**).
   10% (v/v) glycerol.
   10% (v/v) ethylene glycol (*see* **Note 5**).

14 mM 2-mercaptoethanol (*see* **Note 6**).

0.3 g/L benzamidine–HCl (*see* **Note 7**).

1 mM phenylmethanesulfonyl fluoride (PMSF) (*see* **Note 7**).

5 mM sodium phosphate (*see* **Note 8**).

50 mM NaF (*see* **Note 8**).

5% (w/v) insoluble polyvinylpolypyrrolidone (PVPP).

5. Polytron homogenizer (Kinematica).
6. Nylon meshes (1 mm and 100 μm).
7. Refrigerated centrifuge.
8. Saturated ammonium sulfate solution in 0.1 M Tris–HCl (pH 7.6) at 4 °C.

2.2.4 Microscale Extraction

1. Liquid nitrogen.
2. TissueLyser II™ (Qiagen) or equivalent bead mill apparatus.
3. Zirconia beads (5-mm diameter).
4. 2-mL Eppendorf microtubes.
5. Extraction buffer (*see* **Note 9**).

   50–100 mM Tris–HCl or Hepes–KOH (pH 7.5–7.8) containing protease and phosphatase inhibitor cocktails.
6. Vortex mixer.
7. Microcentrifuge.
8. Apparatus and reagents for Western blotting (*see* **Note 10**).
9. (Optional) saturated ammonium sulfate solution in 0.1 M Tris–HCl (pH 7.6).

2.3 Materials for the Phosphorylation of PEPC by PEPC-PK In Vitro

1. Spin column (Sephadex G-25 or equivalent) (*see* **Note 11**).
2. Dissolving buffer.

   20 mM Tris–HCl, pH 7.6 (*see* **Note 11**).

   2 mM dithiothreitol (DTT) (*see* **Note 12**).
3. Protease inhibitor cocktail (EDTA-free).
4. Microcentrifuge.
5. PEPC (*see* **Note 13**).
6. PEPC-PK (*see* **Note 14**).
7. 10× reaction mixture (*see* **Note 15**).

   10 mM EGTA.

   1 mM ATP.

   20 mM $MgCl_2$.
8. Apparatus and reagents for Western blotting.

2.4 Materials for Immunodetection

2.4.1 Western Blot Analysis

1. Spin column (*see* Subheading 2.3).
2. Dissolving buffer (*see* Subheading 2.3).
3. Microcentrifuge.
4. Apparatus and reagents for sodium dodecyl sulfate–polyacrylamide gel electrophoresis (SDS-PAGE).
5. Apparatus and reagents for blotting.
6. Apparatus and reagents for immunodetection:
   - Polyvinylidenedifluoride (PVDF) membrane.
   - Blocking reagent (*see* **Note 16**).
   - Phosphopeptide/phosphonopeptide and PEPC antibodies.
   - Secondary antibody: horse radish peroxidase (HRP)-conjugated anti-rabbit antibody.
   - Tris-buffered saline containing 0.05% Tween 20 (TBST).
   - Shaker.
   - Apparatus and reagents for detecting HRP.

2.4.2 Dot Blot Analysis

1. Bio-dot apparatus (Bio-Rad) or equivalent.
2. Bovine serum albumin (BSA).
3. Blocking reagent (*see* **Note 16**).
4. Apparatus and reagents for immunodetection (*see* Subheading 2.4.1).

3 Methods

3.1 Selection of the Peptide Sequence for Immunization

To obtain a phosphorylation site-specific antibody for a target phosphoprotein, a phosphopeptide representing part of the protein sequence is used as an antigen. In general, four to six residues are placed at the *N*- and *C*-terminal sides of the target phosphoserine/threonine/tyrosine residue of the protein. A Cys residue is coupled to the *N*-terminus or *C*-terminus of the selected phosphopeptide to enable the efficient conjugation with the carrier protein [e.g., keyhole limpet hemocyanin (KLH)]. The following sequences were chosen for the maize C_4-form of PEPC.

Phosphopeptide: G-C-P-G-E-K-H-H-pS-I-D-A-Q-L-R
(pS denotes phosphoserine)
Nonphosphopeptide: G-C-P-G-E-K-H-H-S-I-D-A-Q-L-R

3.2 Chemical Synthesis of the Phosphopeptide [19]

1. Add piperidine/ NMP (1/4 v/v, 10 mL) to an Fmoc amino acid-preloaded Wang resin or an Fmoc-protected Rink amide resin (0.25 mmol), and vortex the mixture four times for 2.5 min each to remove the Fmoc group.

2. Wash the resin five times with NMP.
3. Dissolve the Fmoc amino acid (1.0 mmol)/ HCTU/ 6-Cl-HOBt (4/4/4 mol eq.) in NMP, after which add DIEA (8 eq.) to the solution and stir the mixture for 1 min.
4. Add the above solution to the resin and vortex the mixture for 30 min to couple the Fmoc amino acid.
5. After filtering the reaction solution, add Ac_2O/HOBt/DIEA to the resin and vortex the mixture for 5 min to cap unreacted amino groups.
6. Wash the resin five times with NMP.
7. Repeat **steps 1–6** to elongate the desired peptide chain.
8. Add piperidine/NMP (1/4 v/v, 10 mL) and vortex the mixture four times for 2.5 min each to remove the Fmoc group.
9. Wash the obtained protected peptide resin successively with NMP, dichloromethane, and diethyl ether, after which dry it under vacuum.
10. Add TFA (20 mL), TIS (0.54 mL), H_2O (0.54 mL), and thiophenol (0.54 mL) to the peptide resin (0.25 mmol) in a round-bottom flask, with stirring at room temperature for 1.5–2 h.
11. Add dry cool diethyl ether (300 mL) to the reaction mixture and collect the precipitate by filtration with a microfilter. Extract the crude phosphopeptide with a proper solvent (e.g., 0.1% TFA aq.) to separate the resin and lyophilize.
12. Purify the crude phosphopeptide by preparative RP-HPLC (CH_3CN/0.1% TFA aq. linear gradient system). Collect the desired pure fractions and lyophilize to obtain the pure phosphopeptide. The phosphopeptide yield is generally 100–200 mg (corresponding to an overall yield of 30–70%).
13. The obtained phosphopeptide should be identified by mass spectrometry and the amino acids should be analyzed following an acid hydrolysis. The purity of the final product should be determined by analytical RP-HPLC.

3.3 Production of a Phosphorylation Site-Specific Antibody Using the Phosphopeptide as an Antigen (e.g., PC4 Antibody)

3.3.1 Conjugation of the Phosphopeptide Antigen to the Carrier Protein

1. Dissolve KLH (Calbiochem, 100 mg) in phosphate-buffered saline (PBS, pH 8.0, 3 mL) and stir the mixture gently.
2. Slowly add EMCS (40 mg) in DMSO (2 mL) to the KLH solution and stir the mixture at room temperature for 3 h.
3. Dialyze the reaction mixture against PBS several times at 4 °C.
4. Add the phosphopeptide antigen (5 mg) in 8 M urea (pH 7, 0.2 mL) to maleimide-activated KLH (15 mg) and stir the mixture at room temperature overnight.
5. Dialyze the reaction mixture against H_2O several times.
6. Lyophilize the obtained phosphopeptide-KLH conjugate.

3.3.2 Immunization

1. Immunize two rabbits with the peptide-KLH conjugate and Freund's adjuvant emulsified in a standard way.
2. Collect blood at an appropriate time and separate the serum by centrifugation.

3.3.3 Precipitation of Antibodies with Ammonium Sulfate

1. Slowly add enough saturated ammonium sulfate solution to the antiserum being stirred gently to bring the final concentration to 40% saturation.
2. Centrifuge the suspension at 2000–3000 × *g* for 30 min.
3. Remove and discard the supernatant.
4. Resuspend the pellet in PBS.
5. Transfer the antibody solution to a dialysis tube and dialyze against PBS (three changes) overnight.

3.3.4 Preparation of Affinity Columns

1. Remove the supernatant of the amino-terminal crosslinked agarose gel (Affi-Gel 102 Gel, Bio-Rad, 4–5 mL) and resuspend in PBS (pH 8.0, 10–15 mL).
2. Add EMCS (30 mg) dissolved in DMSO to the medium.
3. Mix the medium by end-over-end shaking overnight at room temperature.
4. Remove the excess EMCS by washing with H_2O.
5. Add the phosphopeptide antigen (5 mg) or the nonphosphopeptide (5 mg) in PBS to the medium.
6. Mix the medium by end-over-end shaking overnight at room temperature.
7. Remove the excess peptide by washing with H_2O.
8. Transfer the medium to blocking buffer (0.1 M mercaptoethylamine in PBS) to block any remaining active groups and mix by end-over-end shaking overnight at room temperature.
9. Remove excess blocking agent by washing with PBS.
10. Transfer the medium to a column to obtain the phosphopeptide- or nonphosphopeptide-coupled agarose gel column.

3.3.5 Affinity Purification

1. Apply the antibody solution to the nonphosphopeptide-coupled agarose gel column at a low flow rate.
2. Reapply the flow-through fraction several times.
3. Apply the collected flow-through fractions to the phosphopeptide-coupled agarose gel column.
4. Reapply the flow-through fraction several times.
5. Wash the phosphopeptide-coupled agarose gel column with 1 M NaCl in phosphate buffer (pH 7) to eliminate nonspecific adsorption and then equilibrate the column with PBS.

6. Elute the antibodies from the column with 0.1 M glycine–HCl (pH 2.3).
7. Collect the eluate in a flask containing 0.5 M Tris–HCl (pH 7.5) to immediately return the antibodies to neutral conditions.
8. Dialyze against PBS or concentrate the eluate as necessary.
9. The antibodies can be used without replacing the solvent, but if necessary, replace with PBS. Store the stock solution at 2–10 °C for short-term storage or at −20 °C or −80 °C for long-term storage. Avoid a freeze/thaw cycle.

3.3.6 Enzyme-Linked Immunosorbent Assay (ELISA) for Measuring Specificity

The titer of the purified antibody should be determined by ELISA.

1. Conjugate the phosphopeptide or nonphosphopeptide to OVA to decrease the difference between wells.
2. Dissolve OVA (Sigma-Aldrich, 100 mg) in PBS (pH 8.0, 1.5–2 mL) and stir the mixture gently.
3. Slowly add EMCS (15 mg) or GMBS (15 mg) in DMSO (0.5 mL) to the OVA solution and stir the mixture at room temperature for 3 h.
4. Dialyze the reaction mixture against PBS several times at 4 °C.
5. Add the phosphopeptide (2–3 mg) or nonphosphopeptide (2–3 mg) in 8 M urea (pH 7, 0.2 mL) to maleimide-activated OVA (10 mg) and stir the mixture at room temperature overnight.
6. Dilute to 2 mL with H_2O or PBS.
7. Coat half of the wells of a 96-well microtiter plate with 100 μL of the appropriate phosphopeptide-OVA conjugate at a concentration of 20–50 μg/mL in coating buffer (carbonate buffer, pH 9.5).
8. Coat the other half of the wells of the 96-well microtiter plate with 100 μL of the appropriate nonphosphopeptide-OVA conjugate at a concentration of 20–50 μg/mL in coating buffer (carbonate buffer, pH 9.5).
9. Cover the plate and incubate overnight at 4 °C or room temperature.
10. Wash the plate four times with H_2O.
11. Add 100 μL of blocking solution (casein in PBS) to each well. Incubate for several hours (or longer) at 4 °C or room temperature.
12. Remove the blocking solution.
13. Add 100 μL of suitably diluted samples to the relevant wells.
14. Incubate overnight at 4 °C or room temperature.

15. Wash three times with H_2O.
16. Add 100 μL of the horseradish peroxidase-conjugated secondary antibody (appropriately diluted in PBS containing BSA) to each well and incubate for 90 min at room temperature.
17. Wash three times with H_2O.
18. Add 100 μL of ABTS solution (0.4 mg/mL in acetate buffer, 50 mM, pH 4.0) to each well and incubate at room temperature for 15 min.
19. Read absorbance values immediately at 415 nm.
20. Compare the titers for the phosphopeptide and nonphosphopeptide and if there is any antibody for the nonphosphopeptide remaining, apply the antibody solution to the nonphosphopeptide-coupled agarose gel column again.

3.4 Production of a Phosphorylation Site-Specific Antibody Using the Phosphonopeptide as an Antigen (Exemplified by PNB Antibody Production)

A phosphorylation site-specific antibody may also be produced with phosphonopeptide (phosphopeptide mimetic) as an antigen. In this peptide, phosphoserine is replaced with (*S*)-2-amino-4-phosphonobutyric acid (L-Apb) (Fig. 1). Because the phosphoryl group in L-Apb is supposedly resistant to dephosphorylation by endogenous phosphatases during the immunization of rabbits, better antibodies with a higher titer are expected [20]. The following sequence was chosen for the maize C_4-form of PEPC as an antigen.

Phosphonopeptide: G-C-P-G-E-K-H-H-X-I-D-A-Q-L-R
X=L-Apb

3.4.1 Chemical Synthesis of Phosphonopeptide

1. Synthesize the key material, Boc-L-Apb(Me_2) (Fig. 1), as previously described by G. Tong et al. [21].
2. Add 50% TFA/ CH_2Cl_2 to a Boc-Arg(Tos)- PAM resin (0.3 g, 0.15 mmol) and vortex the mixture two times (3.5 and 16.5 min) to remove the N^{α}-Boc group.
3. Wash the resin five times with CH_2Cl_2, neutralize with 5% DIEA/NMP two times and then wash it five times with NMP.
4. Dissolve the Boc amino acid/DCC/HOBt(13/13/13 mol eq) in NMP, mix the solution for 40 min and filtrate.
5. Add the filtrate to the resin and vortex the mixture for 1 h to couple the Boc amino acid.
6. After filtering the reaction solution, add Ac_2O/HOBt/DIEA to the resin and vortex the mixture for 5 min to cap unreacted amino groups.
7. Wash the resin four times with NMP and five times with CH_2Cl_2.
8. Repeat **steps 2–7** to elongate the peptide chain until the I-D(OHex)-A-Q-L-R(Tos)-resin is prepared.

9. For the coupling of Boc-L-Apb(Me_2), add Boc-L-Apb(Me_2) (1.2 eq), WSCD (1.5 eq) and HOBt (1.2 eq) in DMF to the resin and stir at room temperature for 8 h.
10. Repeat **steps 2-7** to elongate to obtain the desired protected phosphonopeptide-resin.
11. Add 50% TFA/CH_2Cl_2 to the resin and vortex the mixture two times (3.5 and 16.5 min) to remove the N^{α}-Boc group.
12. Wash the resin five times with CH_2Cl_2, neutralize with 5% DIEA/NMP two times, and then five times with NMP.
13. Wash thus obtained desired protected phosphonopeptide-resin successively with NMP, CH_2Cl_2, and diethyl ether, after which dry it under vacuum.
14. Transfer the resin to an HF-reaction vessel (resistant to HF), to which add *p*-cresol (2 mL). Set the vessel to an HF-reaction apparatus, to which add HF (8 mL), and stir the mixture for 60 min at −5 °C to detach the resin support and remove all protected groups except for two methyl groups on Apb.
15. After concentrating the reaction mixture in vacuo at −5 °C, add dry cool diisopropyl ether (200 mL) to the reaction mixture and collect the precipitate by filtration with a microfilter. Extract the crude Apb(Me_2)-phosphonopeptide with 0.1% TFA aq. to separate the resin and lyophilize.
16. Treat the crude Apb(Me_2)-phosphonopeptide with dithiothreitol (5 eq) in aqueous ammonia for 10 min, acidify the mixture with TFA, and then purify it by preparative RP-HPLC (CH_3CN/0.1% TFA aq. linear gradient system).
17. Add TMSOTf-dimethyl sulfide-*p*-cresol-TFA (3.0, 2.3, 1.6, 5.2 mL, respectively) to the purified Apb(Me_2)-phosphonopeptide and stir the mixture for 18.5 h at 0 °C to remove the remained two methyl groups on Apb.
18. Add dry cool diisopropyl ether (300 mL) to the reaction mixture and collect the precipitate by filtration with a microfilter.
19. Purify the crude phosphonopeptide by preparative RP-HPLC (CH_3CN/0.1% TFA aq. linear gradient system). Collect the desired pure fractions and lyophilize to obtain the pure phosphonopeptide as a white amorphous powder.
20. The obtained phosphonopeptide should be identified by mass spectrometry and the amino acids should be analyzed following an acid hydrolysis. The purity of the final product should be determined by analytical RP-HPLC.

The obtained phosphonopeptide is conjugated to the carrier protein for a subsequent immunization of rabbits, after which the antisera are purified by affinity chromatography with a procedure that is almost the same as that used for the phosphopeptide described in Subheading 3.3.

3.5 Methods for the Extraction of Proteins from Maize Leaves (See Notes 2 and 7)

1. Harvest maize leaves. For mature leaves, remove the midribs. Immediately homogenize the leaves according to the following procedure or freeze materials with liquid nitrogen and store at −80 °C.

3.5.1 Large-Scale Extraction

2. Proteins should be extracted at 9 °C or lower within 30 min of harvesting the leaves (*see* **Note 7**). When mature leaves are used immediately for a protein extraction, mince leaves (not frozen) using a kitchen knife. Suspend the minced leaves in 5 volumes (per wet weight) of the ice-cold extraction buffer. Roughly disrupt the leaves with a Waring blender and further homogenize with a polytron homogenizer. When frozen leaves are used as the starting material, pulverize the leaves in liquid nitrogen with a pestle chilled with liquid nitrogen and then suspend the leaf material in the ice-cold extraction buffer for a subsequent homogenization with the polytron (*see* **Notes 17** and **18**).
3. Filter the homogenate through layers of nylon meshes (1 mm and 100 μm, in turn) (*see* **Note 19**), and then centrifuge at 6000 × *g* for 10 min. Fractionate the supernatant with saturated ammonium sulfate in 0.1 M Tris–HCl (pH 7.6) (25–60% saturation). Dissolve the precipitate in the extraction buffer without PVPP. Bring the solution to 60% saturation with the saturated ammonium sulfate solution. Freeze the resulting suspension with liquid nitrogen and store at −80 °C.

3.5.2 Micro-scale Extraction (See Note 20)

1. Prepare the required number of zirconia bead sets in a 2-mL Eppendorf microtube.
2. Weigh detached leaf samples and add them to the microtubes (*see* **Note 21**).
3. Flash-freeze the samples with liquid nitrogen and store them at −80 °C until used.
4. Prechill the TissueLyser II™ adapter tube holder (2 × 24) in a freezer (*see* **Note 22**).
5. Prepare two or more polyacrylamide gels and 3× gel loading buffer (including 6% SDS) for the SDS-PAGE. Prior to the next step, dispense the 3× gel loading buffer into each tube and set gels in an electrophoresis tank (*see* **Note 10**).
6. Set the frozen sample tubes into a chilled holder under freezing conditions provided by liquid nitrogen to avoid melting.
7. Disrupt leaf tissues with the TissueLyser II™ at 18 Hz for 80 s, making sure to prevent the samples from melting.
8. Add 10–20 volumes (per wet weight of sample) of extraction buffer to the powdered samples in tubes, after which quickly mix the solutions with a vortex mixer for 1 s and centrifuge at maximum speed (e.g., 15,000 × *g*) for 1 min.

9. Immediately recover the supernatant and add 2 volumes to 1 volume of 3× gel loading buffer and mix well. Alternatively, fractionate supernatants with ammonium sulfate (25–60% saturation) and store as ammonium sulfate suspensions (60% saturation) at −80 °C until used (*see* **Note 23**).
10. Immediately subject the samples to SDS-PAGE followed by an immunodetection analysis (*see* **Note 10**).

3.6 Method for Phosphorylating PEPC In Vitro

1. When the extracts are stored as an ammonium sulfate suspension, remove the ammonium sulfate from the extracts with a spin column according to the following procedure prior to the phosphorylation reaction (*see* **Note 11**). For the frozen extracts, thaw and centrifuge samples. Dissolve the protein precipitates with a small amount of the dissolving buffer. Desalt the solutions with spin columns preequilibrated with the dissolving buffer, after which add a protease inhibitor cocktail (EDTA-free) to the eluents.
2. Start the phosphorylation reaction by adding the 10× reaction mixture and PEPC-PK [final 0.5 ng/μL [15]] to the PEPC solution (crude or desalted extract) and then incubate at 30 °C for 15 min (*see* **Notes 11–13**, **15**, and **24**).
3. Stop the reaction by adding 1/2 volume of 3× gel loading buffer and complete a Western blot analysis.

3.7 Methods for Immunodetection

For extracts containing ammonium sulfate, desalt samples (*see* Subheading 3.6) prior to the SDS-PAGE.

3.7.1 Western Blot Analysis

1. Complete the SDS-PAGE with dual gels for the subsequent blotting, with one gel used for the phosphopeptide/phosphonopeptide antibody and the other gel used for the PEPC whole protein antibody.
2. After the electrophoresis, transfer the proteins to PVDF membranes according to a semidry or tank method. Block membranes with the blocking reagent at room temperature for 30 min, after which add the primary antibody.
3. After incubating with the primary antibody, add the HRP-conjugated secondary antibody and detect the chemiluminescence with the chemiluminescence system. Quantify the signal intensities derived from the phosphopeptide/phosphonopeptide antibody with a luminoimager and normalize the signals based on the corresponding signals for the PEPC antibody. If a luminoimager is not available, complete a dot blot analysis, which is a useful semiquantitation method.

3.7.2 Dot Blot Analysis (See Notes 25 and 26)

The dot blot analysis and the Western blot analysis involve the same method except for the blotting. Precipitate the proteins in the ammonium sulfate suspension and dissolve with PBS (*see* **Note 27**). Serially dilute the extracts with PBS containing 1 μg/mL BSA and blot samples onto a PVDF membrane using the Bio-Dot apparatus. Prepare two sheets for the blotted membrane, with one used for detecting phosphorylated PEPC with the phosphopeptide/phosphonopeptide antibody and the other used for detecting the PEPC protein with an antibody as described for the Western blot analysis.

A luminoimager can be used to measure the signal intensities. The chemiluminescence can also be detected with X-ray film. If a luminoimager and a dark room are not available, detect signals with a colorimetric method involving 0.5 mg/mL 3,3′-diaminobenzidine and 0.009% (v/v) H_2O_2 in PBS.

3.8 Evaluation and Application of Phosphorylation-Specific Antibodies

3.8.1 Specificity of the Recognition by the Antibodies

The amino acid sequence of the synthetic oligopeptide used to immunize rabbits is based on the maize C_4-form of PEPC (*Zea mays*). The sequence is from the 9th to the 21st residue (-PGEKHHSIDAQLR-), in which a phosphorylated Ser is located at the 15th position from the *N*-terminus (Fig. 2a). Two amino acid residues, Gly (G) Cys (C), are added to the *N*-terminus of this phosphopeptide for the conjugation to KLH (*see* Subheading 3.3.1). The peptide is then used as an antigen (Fig. 2a).

3.8.2 Anti-phosphopeptide (PC4) Antibody

The PC4 antibody specifically recognizes the phosphorylated wild-type C_4-form of the maize PEPC (ZmC4PEPC) (Fig. 2b). The fact that a mutant PEPC (S15D), in which the negatively charged Asp replaces the phospho-Ser^{15} [9], is not detected by the antibody (Fig. 2b) confirms that the phosphoryl group is indispensable for the antibody recognition. The antibody also reacts with the extract prepared from sorghum (*Sorghum bicolor*) leaves (data not shown); however, it does not recognize the maize root-form of PEPC produced in *Escherichia coli* and phosphorylated in vitro (data not shown). The corresponding amino acid sequences of the sorghum C_4-form and maize root-form PEPCs are ASERHHSIDAQLR and MPERHQSIDAQLR, respectively [22, 23]. Thus, Lys^{12} is not required for the antibody recognition, whereas His^{14} is. Specifically, two consecutive His residues and the adjacent highly conserved sequence, pSIDAQLR, are believed to form an epitope. Consistent with this idea, a mutant PEPC (K12N), in which Lys^{12} is replaced by Asn (N), also produces a clear signal only when it is phosphorylated (Fig. 2b, [11]). The corresponding amino acid sequence in the C_3-form of the maize PEPC is KMERLSSIDAQLR (Fig. 2a, [24]). Taken together, we concluded that the PC4 antibody specifically detects the phosphorylated C_4-form of the maize PEPC.

3.8.3 Anti-phosphonopeptide (PNB) Antibody

The C_4-form of the PEPC from a C_4-photosynthetic flaveria species (*Flaveria bidentis*) is also phosphorylated diurnally. The C_4-form of the *F. bidentis* PEPC has Leu (L) and Ala (A) residues instead of the two consecutive His (HH) residues that are in the C_4-form of the maize PEPC (Fig. 2a). An analysis of the extracts

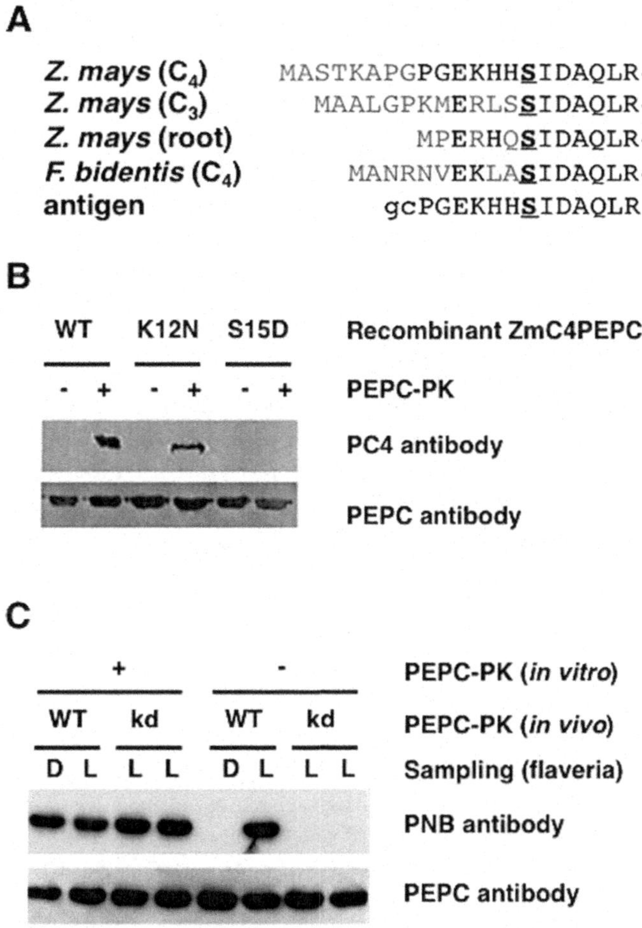

Fig. 2 The specificities of two antibodies. (**a**) *N*-terminal amino acid sequences of three isoforms (C_4-, C_3-, and root-forms) of maize (*Z. mays*) and the C_4-form of Flaveria (*F. bidentis*) PEPCs in which the Ser residues (underlined) involved in regulatory phosphorylation are indicated. The amino acid sequence of the peptide used as an antigen is also provided. The sequence from Pro^9 to Arg^{21} was taken from the C_4-form of the maize PEPC. The Gly and Cys residues (lowercase letters) were attached to the *N*-terminal side for the crosslinking with carrier proteins. The amino acid residues absent in the antigen sequence are in gray. (**b**) The PC4 antibody specifically recognized the phosphorylated C_4-form of the maize PEPC, even when Lys^{12} was replaced with Asn (K12N). Using the partially purified PEPC-PK, the wild-type (WT) and the K12N and S15D recombinant C_4-forms of the maize PEPC (ZmC4PEPC) were phosphorylated in vitro (+), followed by immunodetection. (−) refers to the negative control (i.e., no PEPC-PK). (**c**) Missing signals for the PNB antibody in the transgenic Flaveria plants lacking PEPC-PK. The results indicate that PEPC-PK contributes to the phosphorylation of PEPC in vivo. Leaves of two independent transgenic lines of Flaveria in which the PEPC-PK gene is knocked down (kd), and wild-type (WT) plants were collected at midnight in darkness (D) or at midday under light (L). The extracts were rephosphorylated in vitro with (+) or without (−) recombinant Flaveria PEPC-PK, followed by the immunodetection involving the PNB and PEPC antibodies. The Western blots were reprinted with permission from refs. [11] (**b**) and [15] (**c**) after modifications

from leaves incubated under light and dark conditions revealed that the PNB antibody reacts with the PEPC phosphorylated in vitro or extracted from the illuminated leaves of wild-type flaveria (Fig. 2c) [15]. These results suggest that the antibody specifically recognizes a phosphorylated Ser residue, and the pSIDAQLR sequence is sufficient for the recognition. Because the pSIDAQLR sequence is highly conserved in the PEPCs of diverse vascular plant species, the PNB antibody is useful for detecting phosphorylated PEPCs from various plant species (*see* **Note 28**).

3.9 Examples of the Application of the Two Generated Antibodies

3.9.1 Anti-phosphopeptide (PC4) Antibody

On the basis of its specificity, the PC4 antibody can be used to investigate the dynamic phosphorylation of the C_4-form of the maize PEPC in response to environmental cues. The diurnal changes in the phosphorylation of the C_4-form of the maize PEPC have been confirmed by Western (Fig. 3a) and dot blot (Fig. 3b) analyses [11]. The dot blot results indicate that the PEPC is highly phosphorylated during the day, but is almost dephosphorylated at night.

3.9.2 Anti-phosphonopeptide (PNB) Antibody

Although regulatory phosphorylation was thought to be catalyzed by a unique protein kinase, PEPC-PK [5, 6, 25, 26], the physiological significance of PEPC phosphorylation remained unclear. Additionally, other candidate protein kinases were revealed to phosphorylate PEPC in vitro [27]. The PNB antibody was used to demonstrate that PEPC is not phosphorylated in transgenic flaveria plants lacking PEPC-PK, implying PEPC-PK contributes to the phosphorylation of the C_4-form of PEPC in vivo, at least under normal growth conditions (Fig. 2c) [15]. One of the criticisms of such a study is that relatively low signals for the PNB antibody are not due to dephosphorylation, but are caused by the possible truncation of the *N*-terminus of PEPC. To examine this possibility, the PEPC in crude extracts were phosphorylated by PEPC-PK in vitro, which increased the intensity of the PNB antibody signals, thereby confirming the *N*-terminus of PEPC is intact (Fig. 2c).

4 Notes

1. Although the methods introduced here were established based on *Zea mays* inbred line H84, they may essentially be applicable to other materials.
2. The large-scale extraction method has worked well for decades and is useful for various applications, including immunoassays, enzyme activity assays, and analyses of large samples (e.g., 0.5 kg leaves or more). The micro-scale extraction method enables the rapid handling of many samples and small sample amounts.
3. Buffer components were considered regarding PEPC stability and the cost and availability of reagents decades ago. The buf-

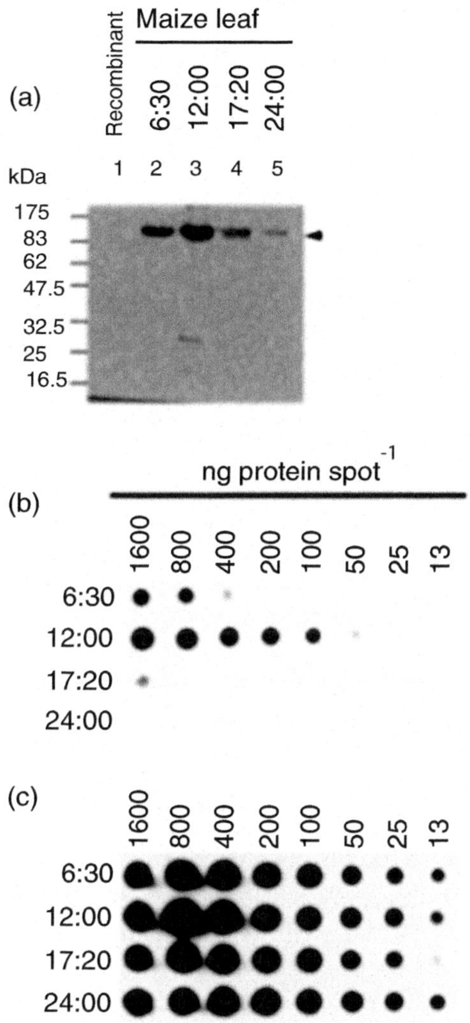

Fig. 3 Example of a semiquantitative dot blot assay involving the PC4 antibody. In a previous study [11], the diurnal phosphorylation of the C_4-form of the maize PEPC was monitored. The following description is directly from the cited reference, and has been reprinted with permission. Samples were prepared from 13-week-old (17 July) maize plants grown in the field and subjected to Western (**a**) and dot blot (**b**, **c**) analyses. Samples of each crude extract (16 μg protein) were loaded in lanes 2–5 and 1 μg of recombinant wild-type C_4-form PEPC of maize was loaded in lane 1 as a control (**a**). The amount of loaded protein is indicated above the dot blots (**b**, **c**). Each membrane was blocked with casein and probed with PC4 (**a**, **b**) or PEPC (**c**) antibodies. Sunrise occurred at about 05:00 h

fers may be altered based on the current research environment. Additionally, the buffer composition depends on each plant material being studied. For example, soluble polyvinylpyrrolidone is highly effective for the extraction of the PEPC from *Eleocharis vivipara* [28], whereas its advantages over other materials for the extraction of PEPC from maize have not been shown.

4. If the extract is used directly in the phosphorylation reaction, take care that EDTA is carried over.
5. Ethylene glycol stabilizes PEPC [29].
6. Dithiothreitol (DTT) is an alternative option. If it is used, it should be freshly added for a final concentration of 1–10 mM.
7. Because the *N*-terminal region containing the phosphorylation site of maize PEPC is readily truncated, precautions should be taken to prevent proteolysis. For example, benzamidine–HCl and PMSF can be used as protease inhibitors. Alternatively, commercially available protease inhibitor cocktails can be used. A 100 mM stock PMSF solution prepared in ethanol should be stored at −20 °C, and then added fresh to the working solution because its half-life in aqueous solutions is 30 min. However, the *N*-terminal of most of the PEPCs in the crude extract is proteolyzed following a 1-h incubation at room temperature, even in the presence of protease inhibitors. Consequently, in addition to the use of protease inhibitors and the completion of the experiment at low temperatures, samples should be processed as quickly as possible. Moreover, PEPC is considerably more stable in an ammonium sulfate suspension.
8. Precautions should also be taken to prevent the dephosphorylation by various phosphatases liberated after cells are disrupted. Phosphate and fluoride are inhibitory to phosphatases. Alternatively, commercially available phosphatase inhibitor cocktails may be used.
9. Precautions should be taken to prevent proteolysis (*see* **Note 7**). Denaturing detergents, such as SDS, should not be used because PEPCs are apparently more susceptible to proteolysis if the extraction buffer contains 0.1% SDS.
10. In this method, samples should be quickly loaded onto gels. Therefore, the SDS-PAGE system should be prepared in advance, if ammonium sulfate fractionation is not performed.
11. The PEPC-PK activity in vitro is sensitive to high salt concentrations [6]. Specifically, the addition of 50 and 100 mM NaCl to the reaction mixture decreases the activity to 50% and 25%, respectively. The optimum pH range for PEPC-PK is 7–8.
12. The PEPC-PK activity is quite sensitive to oxidization [6].
13. Because of its relatively large volume, the PEPC solution [pH, salts, Mg^{2+} (EDTA), reducing agents, and protease inhibitors] affects the composition of the reaction solution. Therefore, the composition of the PEPC solution should be noted. Additionally, PEPC-PK does not phosphorylate heat-denatured PEPC polypeptides [6].
14. Because the C_4-form of the maize PEPC can be phosphorylated at the regulatory phosphorylation site Ser^{15} by mamma-

lian cyclic AMP-dependent protein kinase (PK-A) [30, 31], commercially available PK-A may be alternatively used, although the buffer components and reaction conditions should be reconsidered.

15. The optimum concentrations of ATP and $MgCl_2$ for the PEPC-PK-catalyzed reaction are 0.1 mM and 1–3 mM, respectively.
16. Although skim milk works well, commercially available blocking reagents (e.g., Blocking ONE, Nacalai) are also useful.
17. It is critical that the disrupted leaves remain frozen before they are added to the liquid extraction buffer.
18. If the weight of the leaf material is 10 g or less, the procedure involving liquid nitrogen (described later) is appropriate.
19. For relatively small sample volumes, a single layer (100 μm) is sufficient or the filtration step may be eliminated because the aim of filtration is to remove excess debris.
20. It is critical that the procedure (**steps 6–10**) is completed within 30 min (*see* **Note 7**). Ten samples need approximately 15 min. Therefore, no more than 20 samples should be processed simultaneously.
21. If necessary, divide each plant material prior to storage (e.g., for other applications or reproducibility tests). The protein in the crude liquid extract is unstable. In contrast, the protein is more stable in frozen intact cells stored in a deep freezer.
22. The sudden freezing of a warm (at room temperature) adapter tube holder in liquid nitrogen tends to result in cracks.
23. For the direct analysis by SDS-PAGE (procedure described earlier), the supernatant recovery volume should be twice the volume of the 3× gel loading buffer. The remaining solution (not subjected to SDS-PAGE) is discarded. In contrast, for the ammonium sulfate fractionation (procedure described later), the supernatant recovery volume is not restricted. In fact, a large volume is ideal because the ammonium sulfate precipitation procedure concentrates and stabilizes proteins.
24. The optimum temperature for PEPC-PK is 20–30 °C. Lower temperatures may be appropriate (50% activity at 4 °C), but higher temperatures should be avoided (0% activity at 40 °C).
25. Samples fractionated by ammonium sulfate are favorable for dot blot analysis.
26. Because the dot blot assay provides no information regarding protein size, independently complete the Western blot analysis, if necessary.
27. Desalting is not required for dot blotting.

28. The amino acid sequence around the phosphorylation site, which is conserved among most PEPCs of vascular plants, is (E/D)(K/R)xySIDAQLR, where xy denotes a variety of two amino acid residues [2, 4, 9]. Although the affinity of the PNB antibody is highest for the C_4-form of the maize PEPC, the antibody can also bind to the phosphorylated PEPCs of *Flaveria bidentis* [15] and the following plant species: *Amaranthus cruentus* [16], *Amaranthus hypochondriacus* [17], *Oryza sativa* [18], *Leersia japonica* [18], *Isachne globosa* [18], *Panicum bisulcatum* [18], *Triticum aestivum* [18], *Hordeum vulgare* [18], *Helianthus annuus* [18], *Cucumis sativus* [18], *Cucurbita maxima* [18], *Nicotiana tabacum* [18], *Solanum melongena* [18], *Pisum sativum* [18], *Glycine max* [18], *Arabidopsis thaliana* [18], *Agapanthus africanus* [18], *Lotus japonicus* [18], and *Medicago truncatula* [18]. Thus, the PNB antibody seems to have a wider applicability than the PC4 antibody. Future studies should assess whether the antigenicity of the phosphono residue is far stronger than that of the phospho residue. If it is, the contributions of the variable amino acid residues denoted earlier as x and y in the epitope may be relatively small. A previous study revealed that the production of a phosphorylation-specific antibody for p53 was successful when a phosphonopeptide was used, but not when a phosphopeptide was used [20]. Thus, phosphonopeptides are potentially applicable as alternative antigens for producing phosphorylation-specific antibodies.

Acknowledgments

We thank Edanz Group (www.edanzediting.com/ac) for editing a draft of the manuscript.

References

1. Chollet R, Vidal J, O'Leary MH (1996) Phosphoenolpyruvate carboxylase: a ubiquitous, highly regulated enzyme in plants. Annu Rev Plant Physiol Plant Mol Biol 47:273–298
2. Izui K, Matsumura H, Furumoto T et al (2004) Phosphoenolpyruvate carboxylase: a new era of structural biology. Annu Rev Plant Biol 55:69–84
3. O'Leary B, Park J, Plaxton WC (2011) The remarkable diversity of plant PEPC (phosphoenolpyruvate carboxylase): recent insights into the physiological functions and post-translational controls of non-photosynthetic PEPCs. Biochem J 436:15–34
4. Vidal J, Chollet R (1997) Regulatory phosphorylation of C_4 PEP carboxylase. Trends Plant Sci 2:230–237
5. Hartwell J, Gill A, Nimmo GA et al (1999) Phosphoenolpyruvate carboxylase kinase is a novel protein kinase regulated at the level of expression. Plant J 20:333–342
6. Saze H, Ueno Y, Hisabori T et al (2001) Thioredoxin-mediated reductive activation of a protein kinase for the regulatory phosphorylation of C4-form phosphoenolpyruvate carboxylase from maize. Plant Cell Physiol 42: 1295–1302

7. Aldous SH, Weise SE, Sharkey TD et al (2014) Evolution of the phosphoenolpyruvate carboxylase protein kinase family in C_3 and C_4 *Flaveria* spp. Plant Physiol 165(3):1076–1091
8. Dong L-Y, Ermolova NV, Chollet R (2001) Partial purification and biochemical characterization of a heteromeric protein phosphatase 2A holoenzyme from maize (*Zea mays* L.) leaves that dephosphorylates C_4 phosphoenolpyruvate carboxylase. Planta 213:379–389
9. Ueno Y, Hata S, Izui K (1997) Regulatory phosphorylation of plant phosphoenolpyruvate carboxylase: role of a conserved basic residue upstream of the phosphorylation site. FEBS Lett 417:57–60
10. Takahashi-Terada A, Kotera M, Ohshima K et al (2005) Maize phosphoenolpyruvate carboxylase. Mutations at the putative binding site for glucose 6-phosphate caused desensitization and abolished responsiveness to regulatory phosphorylation. J Biol Chem 280:11798–11806
11. Ueno Y, Imanari E, Emura J et al (2000) Immunological analysis of the phosphorylation state of maize C_4-form phosphoenolpyruvate carboxylase with specific antibodies raised against a synthetic phosphorylated peptide. Plant J 21:17–26
12. Fukayama H, Hatch MD, Tamai T et al (2003) Activity regulation and physiological impacts of maize C_4-specific phosphoenolpyruvate carboxylase overproduced in transgenic rice plants. Photosynth Res 77:227–239
13. Fukayama H, Tamai T, Taniguchi Y et al (2006) Characterization and functional analysis of phosphoenolpyruvate carboxylase kinase genes in rice. Plant J 47:258–268
14. Taniguchi Y, Ohkawa H, Masumoto C et al (2008) Overproduction of C_4 photosynthetic enzymes in transgenic rice plants: an approach to introduce the C_4-like photosynthetic pathway into rice. J Exp Bot 59:1799–1809
15. Furumoto T, Izui K, Quinn V et al (2007) Phosphorylation of phosphoenolpyruvate carboxylase is not essential for high photosynthetic rates in the C_4 species *Flaveria bidentis*. Plant Physiol 144:1936–1945
16. Tazoe Y, Hanba YT, Furumoto T et al (2008) Relationships between quantum yield for CO_2 assimilation, activity of key enzymes and CO_2 leakiness in *Amaranthus cruentus*, a C_4 dicot, grown in high or low light. Plant Cell Physiol 49:19–29
17. Avasthi UK, Izui K, Raghavendra AS (2011) Interplay of light and temperature during the *in planta* modulation of C_4 phosphoenolpyruvate carboxylase from the leaves of *Amaranthus hypochondriacus* L.: diurnal and seasonal effects manifested at molecular levels. J Exp Bot 62:1017–1026
18. Fukayama H, Fujiwara N, Hatanaka T et al (2014) Nocturnal phosphorylation of phosphoenolpyruvate carboxylase in the leaves of hygrophytic C3 monocots. Biosci Biotechnol Biochem 78:609–613
19. Wakamiya T, Togashi R, Nishida T et al (1997) Synthetic study of phosphopeptides related to heat shock protein HSP27. Bioorg Med Chem 5:135–145
20. Higashimoto Y, Saito S, Tong XH et al (2000) Human p53 is phosphorylated on serines 6 and 9 in response to DNA damage-inducing agents. J Biol Chem 275:23199–23203
21. Tong G, Perich JW, Johns RB (1992) The improved synthesis of Boc-Abu(PO_3Me_2)-OH and its use for the facile synthesis of Glu-Abu(P)-Leu. Aust J Chem 45:1225–1240
22. Duff SM, Lepiniec L, Crétin C et al (1993) An engineered change in the L-malate sensitivity of a site-directed mutant of sorghum phosphoenolpyruvate carboxylase: the effect of sequential mutagenesis and S-carboxymethylation at position 8. Arch Biochem Biophys 306:272–276
23. Dong LY, Masuda T, Kawamura T et al (1998) Cloning, expression, and characterization of a root-form phosphoenolpyruvate carboxylase from *Zea mays*: comparison with the C_4-form enzyme. Plant Cell Physiol 39: 865–873
24. Kawamura T, Shigesada K, Toh H et al (1992) Molecular evolution of phosphoenolpyruvate carboxylase for C_4 photosynthesis in maize: comparison of its cDNA sequence with a newly isolated cDNA encoding an isozyme involved in the anaplerotic function. J Biochem 112:147–154
25. Wang YH, Chollet R (1993) Partial purification and characterization of phosphoenolpyruvate carboxylase protein-serine kinase from illuminated maize leaves. Arch Biochem Biophys 304:496–502
26. Tsuchida Y, Furumoto T, Izumida A et al (2001) Phosphoenolpyruvate carboxylase kinase involved in C_4 photosynthesis in *Flaveria trinervia*: cDNA cloning and characterization. FEBS Lett 507:318–332
27. Ogawa N, Okumura S, Izui K (1992) Ca^{2+}-dependent protein kinase phosphorylates phosphoenolpyruvate carboxylase in maize. FEBS Lett 302:86–88
28. Nagamatsu H, Sakagami A, Yamazaki Y, et al (2010) Development of a highly efficient method for extraction of enzyme proteins from plant materials by the use of skim milk as an

assisting agent: a case of PEPC from *Eleocharis vivipara*. Memoirs of the Faculty of BOST of Kinki University 25:7–16 (in Japanese with English summary)

29. Ogawa N, Kai T, Yabuta N et al (1997) Phosphoenolpyruvate carboxylase of maize leaves: an improved method for purification and reduction of the inhibitory effect of malate by ethylene glycol and bicarbonate. Plant Cell Physiol 38:76–80
30. Terada K, Kai T, Okuno S et al (1990) Maize leaf phosphoenolpyruvate carboxylase: phosphorylation of Ser^{15} with a mammalian cyclic AMP-dependent protein kinase diminishes sensitivity to inhibition by malate. FEBS Lett 259:241–244
31. Jiao JA, Chollet R (1990) Regulatory phosphorylation of Serine-15 in maize phosphoenolpyruvate carboxylase by a C_4-leaf protein-serine kinase. Arch Biochem Biophys 283:300–305

Index

Luis M. Vaschetto (ed.), *Cereal Genomics: Methods and Protocols*, Methods in Molecular Biology, vol. 2072,
https://doi.org/10.1007/978-1-4939-9865-4, © Springer Science+Business Media, LLC, part of Springer Nature 2020

D

E

F

G

T

W

Z

Printed by Printforce, the Netherlands